Springer Texts in Business and Economics

Springer Texts in Business and Economics (STBE) delivers high-quality instructional content for undergraduates and graduates in all areas of Business/Management Science and Economics. The series is comprised of self-contained books with a broad and comprehensive coverage that are suitable for class as well as for individual self-study. All texts are authored by established experts in their fields and offer a solid methodological background, often accompanied by problems and exercises.

Pete Bettinger

Forest Harvest Scheduling

From Linear Programming to Heuristic Search

Springer

Pete Bettinger
School of Forestry and Natural Resources
University of Georgia
Athens, GA, USA

ISSN 2192-4333 ISSN 2192-4341 (electronic)
Springer Texts in Business and Economics
ISBN 978-3-031-89431-2 ISBN 978-3-031-89432-9 (eBook)
https://doi.org/10.1007/978-3-031-89432-9

This Springer imprint is published by the registered company Springer Nature Switzerland AG
The registered company address is: Gewerbestrasse 11, 6330 Cham, Switzerland

This book is dedicated to the Chesterfield County, Virginia, public school system. I might not have been a top student, but your schools equipped me well for the challenges I have encountered.

Preface

Forest harvest scheduling is an area of work that requires a person to have some understanding of mathematics, geography, forest inventory, and forest policies that all act in concert to guide the implementation of land management activities. Further, a broad understanding of forestry and natural resource management is helpful, as well as an acknowledgment that when working with other people, particularly those who develop or provide data, one needs to negotiate well. The end result of a forest harvest scheduling effort are products and information that guide forest and land management decisions. Many public agencies (federal, state, county, city); privately owned groups (individuals, investment organizations, churches, and others); publicly owned companies (vertically integrated, real estate investment organizations, and others); and nongovernmental organizations (NGOs) develop forest harvest schedules to help them understand the possible outcomes of management (present and future) and to communicate a narrative that suggests sustainability may be achieved by following the guidance provided by the harvest schedule. Although there are many aspects of work that a forest planner regularly conducts, this text focuses on the development, through mathematical means, of a feasible and efficient plan of action for forest managers to follow. Oftentimes it is referred to as *harvest scheduling*, but in a broader sense it might alternatively be called *management activity scheduling*. Chapter 1 provides an overview of a number of forestry concepts that are important to harvest scheduling. Concepts that arise in subsequent chapters describe the process of harvest scheduling through different mathematical processes and algorithms.

Linear programming is a method of solving problems that are described with an objective function and a set of constraints, where each is represented by a linear equation. The problem, in other words, is represented by equations free of non-linear constraints and integer assignment requirements. In Chap. 2 we describe how forest harvest scheduling efforts are able to employ linear programming if the model system can be described in a manner consistent with the four assumptions of these types of problems (divisibility, certainty, proportionality, additivity). Often, linear programming is employed to examine strategic forest planning alternatives that span long time frames and large areas, but these conditions are not necessary to use the problem-solving method.

Some real forest management problems require integer value (yes/no or otherwise) assignments to potential decisions. For example, when a stand of trees is being considered for a harvest activity, often the activity will be applied for the entire stand of trees (in practice) rather than some arbitrary portion of the area. In Chap. 3 we describe how mixed integer programming is a solution process that is very similar to linear programming in design, yet assumptions regarding the values some decision variables can be assigned may be different. This chapter describes the general application of common mathematical (exact) methods for solving integer or mixed integer programming problems, such as the branch and bound and cutting plane processes.

Goal programming is a special extension of linear and mixed integer programming whereby deviations from goals are minimized within an objective function. In Chap. 4 we describe these types of management problems, their goals (wood production, wildlife, aesthetics, structural conditions, economic value, etc.), and the desired level of outcomes that need to be defined prior to the application of a method to solve the problem. One distinct difference between goal programming and linear or mixed integer programming is the ability of goal programming to recognize different types of management outcomes in the objective function. Further, the outcomes can be normalized if their range of measurements is different. Even further, weights can be devised to force the achievement of certain goals during certain periods of time at the expense, perhaps, of other goals.

Chapter 5 introduces binary search, a simulation process that involves iterative attempts to locate a solution to a problem. From forest harvest scheduling, the solution is often characterized as a sustainable harvest level for each time period within a time horizon. Simulating potential sustainable harvest levels that are feasible and infeasible results in lower and upper bounds on these harvest levels. An attempt is then made to assess the sustainability of a harvest level midway between these. The outcome of this attempt produces a new upper or lower bound, and a subsequent attempt uses this, along with the previous applicable upper or lower bound, to assess the sustainability of a harvest level midway between these. The process continues until the gap between the upper and lower bound is small enough to convince a reasonable person that future attempts (simulations) are not necessary. The last feasible sustainable harvest level is then reported as the solution to the problem.

In Chap. 6 we describe how one can hope to maximize the objective function value of a problem by only making decisions (moves through the solution space) that improve the value of the function, through hill-climbing search processes. One can also hope to minimize the objective function value of a problem by only making decisions that reduce the value of the function. One distinction between this type of search process and others that follow in this book is the strict adherence to the acceptance of moves or actions in the correct direction.

Random search, as presented in Chap. 7 of this book, refers to the ability of a search process to either (1) modify randomly one feasible solution with the hope of improving the quality of the solution, or (2) construct a single feasible solution using random selections of management activities. Random search is efficient, and perhaps one of the computationally fastest methods for developing feasible solutions to complex problems. Random search, however, may not be very effective in producing high quality solutions to complex problems simply due to the random nature of the search. For research purposes, it seems that a random search-based starting point within a solution space is necessary to produce what some consider to be independent samples (final solutions) to complex problems using other heuristic methods.

In Chap. 8, we introduce threshold accepting, which might be characterized as a refinement of Monte Carlo simulation whereby random changes to the character of a solution to a problem are proposed and found acceptable (a) if the changes improve the quality of the solution, or (b) if the changes result in a different solution whose quality is not much different than the previous (or best) solution. The acceptance criteria, or the threshold, is described in the same units as the units employed in the objective function of a problem. The manner in which the process decides whether to continue forward with an inferior solution suggests that threshold accepting may be more logical to use and more practical to understand than other heuristic search processes. With tactical enhancements that are now common among other heuristic methods, a threshold accepting search process has been shown to produce high quality results for forest planning problems.

Chapter 9 concerns simulated annealing, a heuristic search process that is very similar to threshold accepting, with one key difference that is associated with the acceptance rule employed when an inferior solution is proposed. As with threshold accepting, simulated annealing might be considered a refinement of Monte Carlo simulation. During the search process, a single solution is maintained, and a random perturbation is proposed as a change to it. Should the proposed change result in a higher quality, feasible solution, it is automatically accepted. However, should the change result in an inferior solution, it may be acceptable based on a decision that varies by (a) how distant the objective function value is from the previous (or best) solution value, and (b) how long the heuristic has been functioning. A probability of acceptance is determined from this information. Some enhancements to the basic structure of a simulated annealing search process allow it to produce high quality results for forest planning problems.

Much like threshold accepting and simulated annealing, in its basic form tabu search is an *s*-metaheuristic, which implies that a current, feasible solution to a problem will be transformed into a new feasible solution to a problem through a move in a local neighborhood. In Chap. 10 we describe how with tabu search the best move from the local neighborhood is selected, regardless of whether the value of the new solution is better or worse than the value of the previous solution. The caveat to this process is that, in general, the move cannot consist of one that was selected recently, in terms of iterations of the model. This restriction preventing the use of recently selected moves represents the taboo (or tabu) nature of the search process. Ordinarily, tabu search requires more time to complete a search, as compared to threshold accepting and simulated annealing, because many different alternatives need to be assessed in the local neighborhood before one of them is selected.

A genetic algorithm search process acts much differently than simulated annealing or tabu search in how it attempts to solve a problem, as it employs information from several harvest schedules to search for the best overall plan. Chapter 11 describes how when employing a genetic algorithm search process, a population of feasible solutions is created and maintained, new members are developed by combining traits of current members of the population. In doing so, the overall population quality should improve as the population evolves. The search process is analogous to the birth and death of a group of biological organisms. New members of the population, child solutions, are provided genetic material (assignments of activities to decision variables) from current members of the population, or parent solutions. As the population evolves, at least one member should approach the maximum potential biological quality (the global optimum solution to a problem).

Chapter 12 provides a summary of some associated, yet important, issues related directly to forest harvest scheduling. Several of the forest harvest scheduling methods described in this book have been used in practice to assist in the development of forest harvest schedules. All of the forest harvest scheduling methods described in this book have been used to explore the impact of potential policy options on the management of forests. The choice of method to be employed is a function of the perceived amount of time, data, and human capital available to apply to the effort. If optimality is desired, the range of harvest scheduling methods to employ may be limited. If the speed of computing is deemed important, again the range of scheduling methods to employ may be limited. Therefore, the choice of a forest harvest scheduling method is important since the processes employed and the quality and timeliness of outcomes that they produce may vary. Some of these methods are also related to artificial intelligence, and thus we briefly suggest the features that might support these relationships. Issues of efficiency, inclusivity, feasibility, optimality, and legitimacy are discussed in this chapter.

For those who may use this text as a reading resource for a university course on forest harvest scheduling, the book concludes with an exam. If readers have paid close attention to the topics in each chapter, the celebration of learning should be a relatively painless event.

Athens, GA, USA Pete Bettinger

Acknowledgments

This book represents a collection of my experiences as a forest planner, teacher, and consultant. Over the course of my career, the roles I accepted (and others that I considered) seemed to have been guided by moves selected through a heuristic search process. Ideally, as an undergraduate student at Virginia Tech I would have sought the optimal path to follow for obtaining a rewarding career in forestry. But like many undergraduate students, I failed in this regard. Opportunities were presented, many were considered, some were accepted and followed (and later discarded), and life moved on. As an undergraduate student I enjoyed learning about dynamic programming from Howard Hoganson and I wish I had paid more attention to William Leuschner's lectures on linear and goal programming. Later, at Procter & Gamble Cellulose I was introduced to binary search, and wanted to explore it carefully, yet I was too busy with other tasks. As I moved forward in life, I was fortunate to have pursued a doctorate at Oregon State University at a time when forest harvest scheduling was a very hot topic. The courses and guidance I received from Doug Brodie and John Sessions inspired me to not only tolerate and embrace computer programming, but also to enjoy the search for mathematical solutions to forest-level problems. During this time, I was mentored by Norm Johnson, one of the original developers of the FORPLAN model. Our work together was rewarding in many respects, and I am grateful that he asked me to contribute to the fourth edition of *Forest Management*. Kevin Boston (University of Arkansas at Monticello) and I were fellow doctoral students at this time. Our collaborations over the years have delved into several issues that concern forest harvest scheduling, including advances in heuristic search methods, validation processes for assessing the quality of heuristic solutions, and integration of wildlife habitat models into heuristic search processes.

Since arriving at the University of Georgia, I have been fortunate to work with some of the most talented people in the field of forest harvest scheduling. Lingbo Dong (Northeast Forestry University) has allowed me to contribute to his work on exploring the use of simulated annealing to examine alternative management directions for forests in northeastern China. Hayati Zengin (Düzce University), Ahmet

Yesil (Istanbul University—Cerrahpaşa), Abdullah Akay (Bursa Technical University), İnci Caglayan (Istanbul University—Cerrahpaşa), Zennure Ucar (İzmir Katip Çelebi University), Ahmet Değermenci (Düzce University), Ramazan Akbulut (General Directorate of Forestry), and Volkan Bektas (General Directorate of Nature Conservation) have been some of the most wonderful hosts during my visits to the Republic of Turkey, and I have been fortunate to be associated with them on some of the fascinating harvest scheduling and precision forestry issues that concern the forests of these lands. Mehmet Demirci (Weyerhaeuser) is another Turkish scientist who allowed me to contribute to his work exploring the use of goal programming for the placement of thinning operations. Young-Hwan Kim (Korea Forest Research Institute) and I explored the use of heuristic search processes for the placement of management activities that might modify wildfire behavior. Others who have helped educate me about international issues regarding forest harvest scheduling include Jan Kašpar and Robert Marušák (Czech Life Sciences University), Can Vatandaşlar (Artvin Çoruh University), and Mengping Tang (Zhejiang Agriculture and Forestry University). Chris Cieszewski (University of Georgia) was also always ready with ideas and inspiration for addressing science, life, religion, and politics. Further, Don Grebner (Mississippi State University), Jacek Siry (University of Georgia), and Krista Merry (University of Georgia) have been fantastic collaborators in our efforts to contribute to forest management and planning and geospatial educational programs through the development of journal articles and books. Former students who have explored interesting aspects related to forest harvest scheduling include Tiffany Li (combined metaheuristics), Angela Tsao (particle swarm optimization), Ramazan Akbulut (linear programming-informed initiation points), Jared Smythe (particle swarm optimization), Yong Shan (particle swarm optimization), Jianping Zhu (raindrop method), Hector Restrepo (theoretical minimum solution), and Marie Lennette (simulated annealing).

The publication of this book was not a planned outcome of recent U.S. Department of Agriculture projects within which I was associated, but certainly discussions with collaborators of these projects (Fred Maier, Khaled Rasheed, Daniel Drummond, Chaz Merritt, and Bahaaeddin Alaila) were beneficial in crafting the final product. These projects included *Harvest Scheduling and Spatial Forest Landscape Planning* (McIntire-Stennis GEOZ 0002), *Promoting Economic Resilience and Sustainability of the Eastern US Forests* (PERSEUS) (NIFA 2023-68012-38992), *Improved Forest Management for Small Forest Landowners Through a New Decision-Making Model* (eYield) (NIFA 2018-68006-28095), and *Online Learning in Applied Forestry* (OLAF) (NIFA 2019-68006-29331).

Although I have worked as a forest planner, conducted research on forest harvest scheduling methods, and discussed the intricacies of search processes with other more intelligent people, I still do not consider myself an expert in all things related to operations research and forest harvest scheduling. Certainly, any mistakes or omissions that arise within this book are mine and mine alone.

Contents

List of Figures

List of Tables

About the Author

Pete Bettinger is the Hargreaves Distinguished Professor in Forest Management at the University of Georgia. He received a bachelor's degree in Forestry and a master's degree in Forest Economics and Management from Virginia Tech, and he received a doctoral degree in Forest Resources from Oregon State University. During his early career, Dr. Bettinger worked for the forest industry in the southern and western United States. After these experiences, he conducted research and teaching activities at Oregon State University and the University of Georgia. In recognition of the quality of his educational program, the United States Department of Agriculture honored Dr. Bettinger with the *Southern Regional Award for Excellence in College and University Teaching in the Food and Agricultural Sciences*. Dr. Bettinger has become a *Fellow* of the Society of American Foresters, and this organization honored him with the *Carl Alwin Schenck Award* for his devotion to forestry education. The Society of American Foresters further honored Dr. Bettinger with the *Award in Forest Science* for distinguished individual research in the quantitative, managerial, or social sciences.

Chapter 1
Introduction to Forest Harvest Scheduling

Learning Objectives

The objectives of this introductory chapter on forest harvest scheduling are to help the reader

- Understand the motivations for conducting a harvest scheduling effort
- Understand the challenges facing the development of a forest harvest schedule
- Understand realistic expectations for the outcomes from a harvest scheduling effort

1 Introduction

In modern times, the guidance provided by forest planners arises from the use of computer systems that integrate data and analytical methods. The outcomes of a forest harvest scheduling effort provide forest landowners and managers with management direction they may need concerning the resources under their control (Lämås et al., 2023). Paraphrasing slightly Bare and Norman (1969),

> *One of the principal functions of a forest manager is the preparation of a schedule of management activities. These schedules, essential to efficient planning, transform the allowable cut decision into a working plan of action. Therefore, the forest manager usually attaches significant importance to their development. If the manager visualizes their scheduling problems in terms of either maximizing or minimizing an objective function subject to a set of constraints, it is possible for them to develop optimum schedules over a fixed planning period. This, however, requires familiarity with mathematical programming procedures.*

P. Bettinger, *Forest Harvest Scheduling*, Springer Texts in Business and Economics,
https://doi.org/10.1007/978-3-031-89432-9_1

1.1 Forest Planning

Forest planning often involves four connected phases: assessment, decision-making, implementation, and monitoring (US Department of Agriculture, Forest Service, 2014). This book concentrates on the decision-making phase, and the quantitative methodology one might use to develop a high-quality plan of action for a forest. The idea of developing a plan of action for a forest is not new. For quite some time, in fact, a consideration of topography, soils, tree growth rates, management activity costs, potential markets for products, and transportation systems have been advised (Amery, 1875; Burrows et al., 1898) when developing a *forest working plan* (as it once was called) or *forest management plan* (the more common term today). Aspects of silviculture, economics, and management policies have also long guided the order of implementation of management activities on forests (Wieslander, 1922). Further, nearly 90 years ago, Steane (1935) called for the development of Tasmanian state forest working plans that emphasized sustainability of forest resources. Many different resources could be considered when developing a forest plan, including land, biota, people, technology, databases, and organizational contributions (e.g., knowledge, budget, facilities). Invariably, forest planning efforts can also acknowledge changes in agricultural practices and urbanization (Chambers Jr., 1977), as land uses become intermixed within a landscape. In many cases today, forest planning efforts focus on how potential forest management activities might affect economic, commodity production, wildlife habitat, and aquatic resource (among others) values. While the idea of forest planning is not new, the approaches we take to analyze or develop a plan of action have evolved in concert with the evolution of computers, software, data, and database systems. With these resources, forest planning efforts today can process information unimaginable to planners a century ago.

A planning team, consisting of one or more people, develops a forest plan. In small organizations and organizations where the objectives, constraints, and public input are limited, only a few people may be necessary to collaborate in the development of a forest plan. These people would possess the mapping, economic, biometric, silvicultural, information technology, and communication skills to produce the guidance desired by both field foresters (where to go and what to do) and upper-level decision-makers (how much will be produced and how much revenue [or costs] will be generated). When public input is employed in the development of a forest plan, forest management decisions are guided by the thoughts of a broad spectrum of people who are associated with a variety of different interest groups, agencies, or organizations, each with their own agenda. For example, during the public comment period (1 month), the United States Forest Service received about 124,000 letters concerning the scope of the analysis conducted in conjunction with the most recent Tongass National Forest (Alaska) plan (US Department of Agriculture, Forest Service, 2016). This is likely an extreme case of public participation in forest planning given the unique characteristics of the forest, and it should be noted that only 250 unique letters were received. Nonetheless, this set of information needed to be summarized and each unique concern addressed prior to finalizing the plan. To assist in attempts to gain public and organizational acceptance of a plan, outcomes and graphics from forest planning efforts allow groups of people to examine the relative differences in alternative policies in an open manner. The tools used and outcomes produced by forest planners may act to counter or complement influence, power, or other influential tactics of stakeholders.

An assessment of alternative management strategies for forested land areas over important, often long, time frames is at the heart of forest planning (Pukkala et al., 1997). Forest planning involves combining forest landowner preferences with forest production potential in ways that inform people of the consequences of alternative management strategies (Pukkala, 2002). Through the scheduling of management activities over space and time, a forest harvest schedule provides practical direction to forest and land managers regarding the type of management activities to apply to each piece of land (Jones et al., 1986). Among other things, a planning process can help one select stands (collections of trees) to harvest in a certain order using a certain harvest method. In this regard, dividing up a forest into pieces of land that would seem best to manage separately is not a new idea (Ström, 1822). These efforts would help one to determine how much wood volume might be produced to satisfy social demands, to determine how much forest resources may be needed to satisfy competing demands (e.g., wildlife, recreation), or perhaps to understand which roads will be needed to gain access to certain places (Epstein et al., 1999).

There are many reasons why a forest landowner or forest organization would pursue a forest planning process. At times, a forest plan may be required by law, as in the case of national forests in the United States. A forest plan may be required to facilitate entry into forest certification programs, and a plan may be a requirement for landowners seeking financial assistance for forest management activities. In other cases, a forest landowner or forestry organization might develop a plan simply to understand the economic returns and impacts of sustainability (however defined) that might arise from managing forests in certain ways. However, forest planning—specifically harvest scheduling, particularly when applied to public lands in the United States—can be a controversial issue (MacLeery, 1979). Although there may be many areas of concern that discourage a forest landowner or forestry organization from developing a forest plan, a commitment (time, money, spirit) would seem necessary to successfully produce a plan.

1.2 Forest Harvest Scheduling

In general, scheduling management activities involves searching for good overall solutions to a resource management problem. The search method employed, the number and types of constraints, and a desire (perhaps) to optimize a mathematical function all contribute to the difficulty in solving resource scheduling problems (Thalman et al., 1991). One might argue that unplanned implementation of forest management activities may reduce the ability of a forest landowner to use their resources wisely and efficiently and may further reduce the opportunity of a forest

Fig. 1.1 A coordinator for a United States public agency reviews a management plan with a forest landowner. Credit: Bob Nichols Bob Nichols/Photo courtesy of USDA Natural Resources Conservation Service (https://commons.wikimedia.org/wiki/File:NRCSFL97008_-_Florida_(1825)(NRCS_Photo_Gallery).tif)

landowner to manage those resources sustainably. The need for efficient allocation of forest resources has prompted society to study and invest in economic and operations research methods that have generally been described as *forest harvest scheduling* or referred to at times as more generally *forest management planning*. A chronological sequence of forest management activities, with emphasis on wood products outcomes and their associated revenues and costs, forms a harvest schedule (Iverson & Alston, 1986). A forest harvest scheduling process may also recognize and assess simultaneously several other economic, ecological, and social issues that are important to a plan of action (Yoshimoto & Brodie, 1994; Çağlayan et al., 2018). This assessment process can be overwhelmingly complex and computationally difficult, thus, over the last 60 years, a number of different methods have been devised in an attempt to overcome these difficulties.

The quest for sound guidance, through the activities suggested in a forest management plan, is just one of many societal problems. *Forest harvest schedules* often suggest the best plan of action to meet objectives of a forest landowner. A management plan can help a forest manager or landowner (Fig. 1.1) understand the effects of planned activities on sustained presence and yield of forest products and other resources and can help a forest manager or landowner evaluate trade-offs even though the models, data, and assumptions employed may contain a concerning level of uncertainty and variability (Locher et al., 2012). For large forests with a complex array of management concerns, the number of potential sets of actions to follow can easily be intractable (unsolvable) by the human mind. Therefore, complex real-life harvest scheduling problems are often described mathematically and solved using sophisticated solution techniques. The main outcome of harvest scheduling is information that acts as a *guide* to the actions that might be conducted in the management of a forest (Gross, 1950). In managing a forest, the entrusted people (foresters and others) would ideally implement a management program as closely as possible to the guidance provided by a forest harvest schedule. However, solutions to complex harvest scheduling problems may be difficult to implement in practice because the model (the problem described by linear programming or some other method) may not have described all of the intricacies of the real-life system (Albers et al., 1986), and unplanned events may occur (changes in markets, natural disturbances) that force adjustments to be considered. Nonetheless, the guidance provided by a forest harvest schedule may be considerably better than no guidance at all.

Certainly, one can develop a forest harvest schedule using *seat of the pants* or *back of the envelope* methods. These approaches for developing a schedule of forest management activities to implement over time are not very sophisticated nor difficult to employ and involve addressing a problem based on one's experience and judgment, rather than addressing a problem using data, computers, and algorithms (technology). At the beginning of a calendar year, a forest landowner might simply identify (write down on paper) the different stands of trees to be managed (thinned, fertilized, etc.) in a certain way, based on their intuition and experience, and perhaps some sense of the state of local markets. Or, a forest manager might sort their different stands of trees by their average age and begin harvesting the trees from the top of the list, skipping of course, their favorite hunting spots. For small forest landowners, this form of planning might suffice, but for others, more than likely, a plan of action that was developed using methods such as these would be, to say it kindly, suboptimal. Other forms of ad hoc problem-solving methods, using improvised decision rules, would also likely lead to suboptimal decisions regarding the management of a forest. To be fair, suboptimal management of forests is not necessarily bad. In fact, since one usually cannot fully describe a real system through a model (e.g., a linear programming formulation), the guidance provided by optimization algorithms likely also results in a suboptimal management recommendation for the real system. Further, some forest landowners are not as eager to efficiently or optimally manage their lands as others might be. As we move forward in this book, we will assume that forest landowners, in general, do care about the economic, ecological, and social contributions of their land to themselves or their families or even society. Therefore, they may desire to be guided by a forest plan that is both efficient (does not waste resources) and near-optimal (the best plan from a set of potential, feasible plans).

In developing a plan of action for a forest, two obvious courses of action could be taken. First, one could randomly schedule harvest activities (or any reasonable management activity, for that matter) to the management units (stands) of a forest. As was suggested earlier, expert judgment, seat-of-the-pants, back-of-the-envelope, and other ad hoc methods may be similar to (but perhaps a little better than) a random schedule. More than likely, however, these methods for developing a harvest schedule will lead to an inefficient course of action for the forest. Second, one could determine the optimal course of action for each management unit (stand) in a forest and use these optimal stand-level decisions as the basis for a forest-level management plan. This approach would seem to maximize the use of each individual unit of a forest. However, when viewing management outcomes at the forest-level, this may not be the best course of action unless the stand-level and forest-level objectives and constraints are complementary (Hann & Bare, 1979; Reed & Errico, 1986). Assume, for example, that the set of optimal solutions for each stand (where net present value, bare land value, mean annual increment, forest rent, stand density conditions, or other concerns are maximized), independent of decisions applied to other stands, suggests that *every stand* in a forest should be harvested in some way immediately. If one goal of a landowner was to obtain a regular revenue from their forest over time, or if one goal of a land manager were to produce a regular amount of timber every year for a mill, then harvesting all of the stands immediately would not be consistent with these other desires. Another set of optimal solutions for each stand may suggest harvesting most of the stands during the last year of the time horizon, and this too could be inconsistent with the forest-level goals. Some may still argue that a forest should be managed by selecting and implementing stand-level optimal management decisions. Should this be the case, either there are no overarching forest-wide issues of concern such as those involving wood-flow or habitat, or they are ignored. Forest-level plans may be informed by very good (if not optimal) stand-level decisions, but likely are not composed exclusively of them. While stand-level optimal solutions are good to know, alternative, suboptimal stand-level decisions would likely be needed when searching for good overall solutions to a resource management problem, so that broader forest-level goals can be addressed. With this in mind, a *Principle of Scale* associated with forest harvest scheduling is proposed:

> *The sum of a set of stand-level optimal management decisions does not result in a forest-level optimal management plan; further, the sum of a set of forest-level optimal management plans does not result in a landscape-level optimal course of action.*

Assumptions related to this Principle include:

1. Each stand of trees has its own character (site quality, age, species composition, density).
2. The process for optimizing the management of each stand (rotation age, thinning regime, density, etc.) is independent of other stands in a forest.
3. Broader wood-flow, habitat, and other concerns are not recognized in the search for each optimal stand-level management regime.
4. Individual stands of trees form a forest.
5. The stands within a forest may not be connected nor form a contiguous area.
6. The management of a forest may be guided by broad wood-flow, habitat, and other concerns that are informed by stand-level management decisions yet can extend beyond the scope of an individual stand.
7. To optimize the forest system (for any objective), suboptimal stand-level decisions will be necessarily selected to adhere to the wood-flow, habitat, and other concerns.

8. Individual forests form a landscape.
9. The forests within a landscape may be highly irregular in shape and may be owned by very diverse people or organizations.
10. The management of a landscape may be guided by very broad wood-flow, habitat, and other societal concerns that may be informed by individual forest management plans yet can extend beyond the scope of an individual forest.
11. To optimize the landscape system (for any objective), suboptimal forest-level management plans will be necessarily selected to address very broad wood-flow, habitat, and other societal concerns.

More broadly, some elements of forest harvest scheduling have been applied to land use planning, for a systematic assessment of the options driven by a need for change or a need to adapt to changing conditions (Sen & Chatterji, 2018), where perhaps one of the main outcomes is the coordination of management activity to serve the best interests of the greater societal good (Zon, 1933). Forest harvest scheduling is generally different than land use planning in the sense that the latter may be driven by, and provide guidance for, developmental (houses, buildings, roads, etc.) and associated commercial concerns (Selman, 1997), while the former provides guidance for management (the location and timing of forest-based activities) and associated sustainability concerns. Further, land use planning often is applied to broad, contiguous areas composed of a multitude of landowners, while harvest scheduling might focus on a smaller area composed of scattered, disconnected parcels, and a single landowner. However, the two ideas can be integrated, as in the case of county-owned lands in northern Wisconsin, where planned activities addressing both forest management and other land management issues are described (Bayfield County Forestry and Parks Committee, 2021).

For better or for worse, the notion of a *landscape* has often been confined to the land that an entity owns or manages, and the plans developed often include schedules of potential management activities (e.g., harvest schedules) that are ignorant of the state of forests or the collections of activities that occur outside of the ownership boundary. Regardless of this limitation, forest-level planning efforts are important when an organization or landowner is interested in understanding the cumulative effects of their management strategies over space and time. Like Bayfield County, some organizations have gone further to take the entire landscape into consideration when developing a forest plan. In these cases of landscape planning, the physical extent of the landscape is usually defined by a watershed or political (e.g., county, province) boundary. And while individual or aggregate ownership boundaries may be maintained within the modeled landscape, projections of management effects may not be confined to a single organization's or landowner's objectives. At this geographic level of planning, some ecological and social goals may be better modeled, as some issues can only be assessed when viewed across larger areas.

In landscape-level planning, strategic and tactical planning processes may be applied using different objectives and constraints for different landowners or landowner groups (e.g., Bettinger et al., 2005), yet the overall goal is to determine the character of a landscape sometime in the future, if each of the individual landowning entities behave in a certain manner. A business-as-usual scenario generally suggests each landowner will operate in the future just as they are operating today, and this case would be informed by current laws and regulations and surveys of landowners regarding their management practices. Alternative scenarios would apply different constraints to the landowner groups, perhaps to assess how more intensive forest management may change the overall character of a landscape, or conversely how the desire to emulate the historical character of a landscape may affect the forest management practices employed and the outcomes (economic, ecological, social) that may be important to decision-makers (Independent Multidisciplinary Science Team, 1999).

Unlike forest-level planning, landscape-level planning involves multiple landowners within a broad geographic area in an attempt to balance the economic, environmental, and social goals of multiple stakeholders (Freeman et al., 2015). However, the landscape approach is hampered by the need to both (a) recognize the objectives and constraints of each landowner or land manager and (b) model cross-ownership impacts on natural resources. It has been suggested that while landscape-level optimization can help one to understand human and vegetative processes and interactions across broad areas, the outcomes can be limited by the various interests and values of the stakeholders (Sayer et al., 2013). Further, landscape planning is challenged by the fact that individual landowners often make individual (rather than collective) decisions, and given the heterogeneity inherent in a managed landscape, some landowners may benefit from alternative methods for managing a landscape while others may face financial burdens; thus, policy design can be politically difficult (Bostedt et al., 2021). While landscape-level concerns may be important moving forward in the twenty-first century, forest management may be more effectively defined and modeled when only one landowner is considered. For these reasons, in this book, we focus on the forest-level perspective, which focuses on the objectives and constraints facing a single landowner or organization. Certainly, other geographical perspectives are important, and the methods we describe may be applied to model systems that are developed for these cases.

1.3 Model Systems

A harvest scheduling search process is applied to a simplification of a real-world system, and it is important that planners and decision-makers understand they are only developing a solution to the *model system*. This can pose an interesting dilemma to both planners and decision-makers, as a harvest scheduling model would ideally be designed in such a way that the resulting real-life management decisions are related (as closely as possible) to the decision variables in the model (Iverson & Alston, 1986). Within the mathematics of processes that seek to optimize model systems, there is often clear evidence allowing one to differentiate between what seems correct (feasible or optimal) and what seems incorrect (infeasible or suboptimal). The capabilities of search processes can confound those who assume that the very best solution to a problem can be located even when evidence suggests otherwise. Fadaei et al. (2009) note that overly ambitious support of the performance of optimization models should be tempered a bit:

> *"Optimization models" are a type of algorithm intended to provide the best possible solution to some problem facing an organization. Where the problem itself is so complex that finding the best possible solution could cost more than the benefit of doing so, the optimization models generally do not attempt to find the best possible solution, but instead seek to find extremely good solutions within reasonable cost and time parameters. This in fact is the more common situation.*

Within real-world systems, decision-makers face complex problems that may not translate well to mathematically solvable and implementable plans of action (French, 2023). The important details of some real system models might be relatively easy to understand, but other real system relationships may not be amenable to description by equation or logic. In essence, the models developed are an analogue to the real system of interest (Cushen, 1955). While providing a set of high quality, alternative forest plans to a decision-maker may be inspired by the problem (the management situation), the selection of a forest management plan by the decision-maker may be inspired by other factors (e.g., personal values) that were not included in the model that was developed (Bellenguez et al., 2023). In any event, building models is not an easy task. There may be times when real system relationships may require thousands or millions of equations to properly address the functional relationships between potential forest management actions and potential outcomes of those actions. The development and assessment of constraints can also be a complex matter when ecological, social, and economic issues are integrated into a forest harvest scheduling problem (Lacerda et al., 2023). One challenge facing forest planners is to address directly the issue of model complexity. In a simplistic sense, one can limit model complexity and solve a relaxed problem rather easily and exactly, or one can increase model complexity and solve approximately the same problem. However, forces outside of a planner's office may influence the issue of model complexity. For example, decision-makers within an organization may desire a plan that arises from a highly complex model, given their belief that the organization has the data, technology, and commitment to accomplish the task, just as they would with any other forest information system issue (Bettinger, 1999). In sum, the development of the model system, through equations or logic, is an important and perhaps time-consuming step in the planning process.

Example

The Plantersville Tract (Fig. 1.2) is, for the purposes of this book, a hypothetical forest in South Carolina that encompasses 2400 acres (971 ha) of land. The forest is divided into 81 different management units (timber stands). In this model of a real-world system, the forest age class distribution for the operable lands (Fig. 1.3) indicates that the Plantersville Tract contains a relatively

(continued)

uneven mix of forest ages across the property. The time horizon is 15 years long, and it is divided into 15 one year long time periods. There are 16 potential assignments for each timber stand. One assignment represents not harvesting a stand, and 15 other assignments represent harvesting a stand once during different years within the time horizon. Two harvest schedules were created: a randomly generated plan and a plan that utilizes optimal stand-level decisions. In both of these cases, only final harvests, site preparation, and planting activities were considered in the model system. Stumpage prices were assumed to be $20 per ton, reforestation costs were assumed to be $300 per acre, and no stands less than 20 years of age were allowed to be scheduled for a final harvest. The alternative rate of return (discount rate) assumed in this analysis was 5%. The objective was to maximize discounted revenue from final harvests less discounted regeneration costs that occur immediately after harvest. Under these assumptions, a randomly generated harvest schedule scheduled about 1717 fewer tons of wood for harvest than the stand-level optimal schedule (Table 1.1) even though the random schedule suggested that some timber volume would be scheduled for harvest every year, while the stand-level optimum schedule did not. The random schedule also suggested in years 4 and 10 a schedule of harvests that may be unreasonable due to the limited size of the harvest areas. Given the limitations of both cases, some additional forest-level constraints may be necessary to ensure that the wood flow and the areas scheduled for harvest are more reasonable for the landowner, even though these considerations were not noted in the original model of the real-world system.

2 The Forest Harvest Scheduling Optimization Problem

The field of operations research has been described as an area of work in which the scientific method can be applied to practical situations, and the outcomes can provide resource managers with a quantitative basis for the decisions they need to make (Cushen, 1955). This area of work, which includes many types of optimization and heuristic search processes, developed in earnest in the mid-1900s due to logistical issues associated with the movement of people and equipment involved in the Second World War (Hillier & Lieberman, 1980). For the field of forestry, the rationale for pursuing the optimization of a real-world system is often based on a desire to maximize a set of benefits associated with some resources, while ensuring the protection of other resources (Forest-Range Task Force, 1972). The ultimate goal of harvest scheduling is to provide the best possible guidance to forest managers, and this is accomplished by examining the decision (or solution) space of a management problem and locating a desirable course of action from amongst a great number of potential courses of action (Pukkala, 2002). The use of operations research techniques to address these problems in forestry began around 1960, and shortly

Fig. 1.2 A map of the age class distribution of forests within the Plantersville Tract. Credit: Pete Bettinger

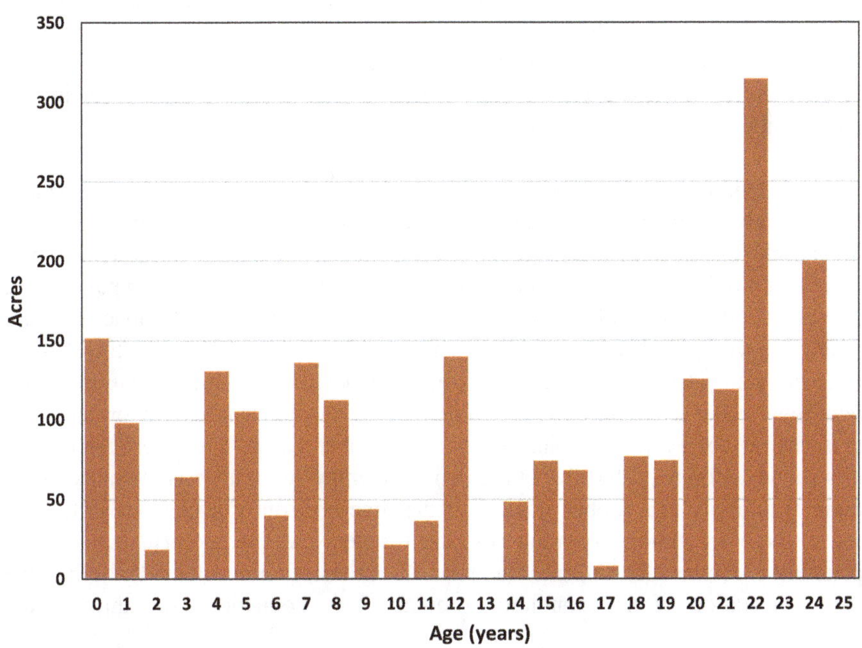

Fig. 1.3 An age class distribution of the forests within the Plantersville Tract. Credit: Pete Bettinger

Table 1.1 Area and volume scheduled for the Plantersville Tract under a random and a stand-level optimum scenario. Credit: Pete Bettinger

	Randomly generated		Stand-level optimum	
Year	Acres scheduled for harvest	Volume scheduled for harvest (tons)	Acres scheduled for harvest	Volume scheduled for harvest (tons)
1	106.1	7572.4	–	–
2	103.0	8175.7	63.6	4893.7
3	22.0	1533.1	192.6	15,977.6
4	1.1	75.9	154.1	11,128.4
5	83.7	6899.2	306.6	25,092.1
6	98.5	5596.1	188.5	15,476.8
7	46.4	3267.4	55.4	4743.9
8	111.6	8559.8	128.3	10,589.5
9	142.1	11,967.8	29.5	2802.8
10	8.0	781.3	–	–
11	75.7	8112.3	–	–
12	164.4	13,960.9	–	–
13	146.6	12,203.2	–	–
14	246.7	19,745.4	143.9	13,254.5
15	478.6	47,308.0	676.8	53,516.4

thereafter some of the early works appeared in literature (Curtis, 1962; Leuschner et al., 1975). These developments coincided with the rapid evolution of computer and database management systems and an increased level of competition for forest land in the United States and elsewhere (Bare et al., 1984). As an example of the efforts expended on investigating the process and application of linear programming, one optimization model, the number of published papers reported in the Web of Science have increased dramatically since it was initially developed in the mid-twentieth century (Fig. 1.4).

Optimization involves the search for, and location of, the very best solution to a problem. Optimization problems common in forestry and natural resource management involve routing, assignment, and resource allocation. *Routing problems* involve the presence of edges (arcs, roads, trails) that form a connected network. A routing problem may involve identifying the shortest path or least cost path from a harvest unit to a sawmill. In a classical sense, these are called capacitated arc problems, where each edge (arc) has a value (benefit, cost, distance, travel time, etc.) and the system desired minimizes these values. In forestry, *assignment problems* involve the selection of an object or choice (management activity or particular prescription) for a location (management unit, timber stand). The object can consist of a single action (e.g., fertilize at age 15) or an entire set of actions (a management regime) to apply to a timber stand over the time horizon of a plan. In a classical sense, these are similar to graph coloring problems. Contemporary spatial forest planning problems also fall into this form of optimization problem. These involve the optimization of an economic or commodity production goal through the

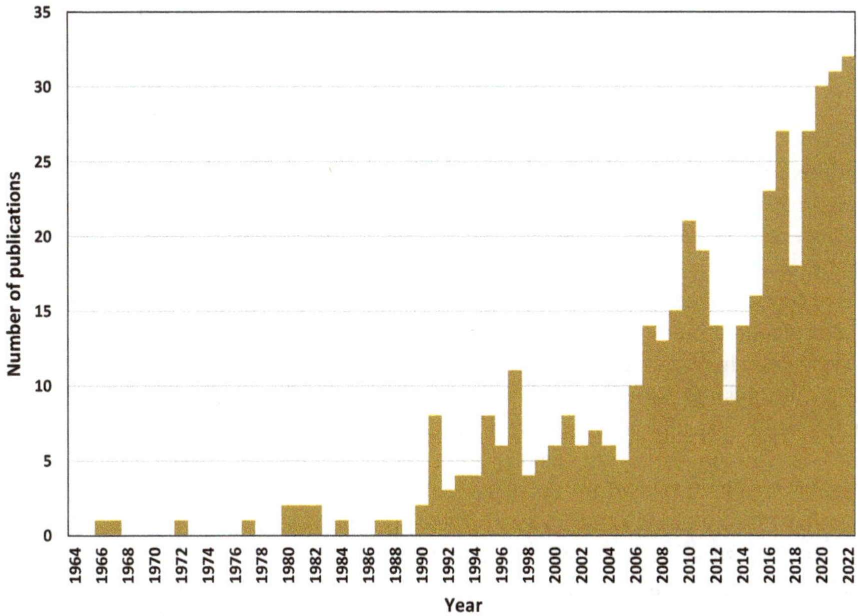

Fig. 1.4 Number of published papers from the Web of Science using the keywords "linear programming" and "forestry." Credit: Pete Bettinger

assignment of a harvest or management regime to each of the locations (timber stands), subject to wood flow, harvest adjacency, and other types of policy constraints. The forest harvest scheduling problems examined in this book are assignment problems. *Resource allocation problems* are similar to assignment problems. In these cases, a set of limited, but available resources are allocated to certain potential management emphases that relate to those resources, with the objective of minimizing or maximizing some objective. Resource-constrained harvest scheduling problems involve assigning time to an activity and ensuring that resource-related constraints are not violated (Thomas & Salhi, 1998).

In developing a forest plan, a forest harvest scheduling problem that emulates a real-life system needs to be described. Initially, one needs to decide upon the *time horizon* or the length of the forest plan in years. Then, one needs to decide upon the specific *time periods* that are recognized within the time horizon. More formally, these have been defined as (U.S. Department of Agriculture, Forest Service, 2014):

> Time horizon: *"The overall time period considered in the planning process that spans all activities covered in the analysis or plan."*
> Time period: *"The time interval within the planning horizon that is used to show incremental changes in yields, costs, effects, and benefits."*

Forest planning often involves assessing management options at three scales: strategic, tactical, and operational. In general, the time horizon assumed is longest at the strategic scale, and it is shortest at the operational scale. A time horizon for a strategic forest plan might be 50–100 years long, while a time horizon for a tactical forest plan might be 10–30 years long, and a time horizon for an operational plan might span only days, weeks, or months. Time periods within a time horizon can be as short as 1–5 years for tactical plans, or 5–10 years for plans with long time horizons. Rarely within the time horizon of forest plan are the time periods of different lengths of time, but it is possible to design a plan with this in mind. Most importantly, regardless of how the time periods are defined, growth projections of forests need to be compatible and consistent with the length of the time period. In addition to these characteristics, the detail utilized to describe a model system is often coarser at the strategic scale and more refined at the operational scale. Strategic forest planning involves understanding sustainability, and the trajectory of a forest or landscape when managed under a set of assumptions for a long period of time. Tactical forest planning involves identifying the set of management actions to be employed physically within a forest and within the time horizon, ideally to satisfy the strategic goals of the landowner (Nelson & Brodie, 1990; Holmgren & Thuresson, 1997). Therefore, the detail required for tactical planning (perhaps spatial relationships amongst small pieces of land) can be more extensive than the detail required for strategic planning. The detail required for operational planning can be even more specific, as the types of problems at this scale focus on the assignment of people, machines, or other items to problems that have a very short time horizon (e.g., the harvest of a stand of trees over the next three weeks). This hierarchical nature of forest planning plays an important role in contemporary forest management activities (Ulvdal et al., 2023) even though the different scales can be linked or integrated when conditions and assumptions are allowed.

Example

A strategic plan. Carle et al. (2020) described the assignment of harvest activities (thinnings, partial harvests, and final harvests) to a 1,198,435-acre (485,000 ha) forest in western Quebec, where 157,000 management units were composed of a variety of mixed species. The time horizon was 150 years long, and the time periods were 5 years long. The objective was to maximize the volume scheduled for harvest during the first 30 years of the harvest schedule. Constraints were imposed to produce a non-declining, relatively even flow of wood products through harvest activities, and to maintain areas of suitable wildlife habitat throughout the 150-year time horizon.

A tactical plan. Demirci and Bettinger (2015) described the assignment of partial harvest activities to a 6464 acre (2616 ha) forest in western Turkey, where 106 management units (compartments) were composed of Anatolian black pine (*Pinus nigra*) and oaks (*Quercus* spp.). The time horizon was 10 years long, and the time periods were 1 year long. The objective was to minimize deviations from a target harvest volume and minimize the average distance between compartments that were scheduled for a partial harvest each

(continued)

year. The constraints limited the number of compartments that could be scheduled for a harvest in each time period.

An operational plan. Mendonça et al. (2022) described the assignment of five different silvicultural activities to a 2224 acre (900 ha) forest in Brazil composed of 32 stands of *Eucalyptus* spp. The time horizon was 40 days long, and the time periods were 1 day long. The objectives were to either minimize the costs, the time to complete the work, or the number of hours needed to complete the work. Silvicultural task sequencing and constraints related to the finite workforce and machines were employed.

An assessment of the goals and constraints of a landowner or land management organization is a necessary first step in understanding a planning problem. These conversations then allow one to develop a model of the system and solve the forest harvest scheduling problem. Clearly, the model is a simplification of the real-world system. But the model should reflect as closely as possible the typical decisions involved, and the costs and benefits of making those decisions. When completed, this model, or *problem formulation*, is subjected to a process that attempts to determine the best course of action, given the desired objective (goal) and the constraints. A relatively straightforward and simplistic objective of forest harvest scheduling may concern producing wood products at a rate equal to the maximum productive capacity of the soil, while safeguarding against soil degradation (Brown, 1894). A more complex objective may concern maximizing various aspects of economic, ecological, and social issues pertaining to the resources within a forest (Merganič et al., 2022).

3 Classical and Contemporary Aspects of Forest Harvest Scheduling

Classical timber production-oriented harvest scheduling efforts were aimed at understanding how different land management alternatives can be selected to produce a minimum sustainable harvest, to achieve economic efficiency, and to make progress toward the development of a regulated age or size class distribution (Marty, 1979). Contemporary forest planning problems extend management concerns more broadly, ecologically, and socially, and may require complex functional relationships (connections between decisions and outcomes) and data analysis methodologies. Some of the classical methods concerning area control (relatively even amounts of areas treated with management activities), volume control (relatively even amounts of volumes scheduled for harvest), or economic control (maximizing or minimizing some economic outcome of a plan) continue to play important roles in contemporary forest planning problems. Roth (1914) once advised that although the time horizon of a forest plan can span many decades, in practice, a plan should not be relied upon for very long and should be revisited at least every 20 years because growth, harvest, mortality, regeneration, market factors, and environmental issues all affect potential harvest levels. In contrast, because they have the data and computing systems to do so, some large forest management organizations revise their contemporary forest plans every year, to account for changes in the resource managed, and to provide current estimates of harvest volumes and associated revenues and costs as market factors evolve.

3.1 A Measurable Objective

As it turns out, just about anything (let us call it G) that can be quantified can be used as an objective in a forest harvest scheduling problem. The caveat is that G must change, vary, or generally be responsive to changes in the variables that reflect decisions that are made. If the decision variables reflect the assignments of management regimes to stands or trees, or more specifically harvests (with associated timing and intensity) to stands of trees, G must either increase or decrease when these decisions are scheduled. If G does not respond or vary when activities are assigned to the landscape, it seems irrelevant to use G as a forest management objective. Some of the quantifiable metrics that can be used to represent G include costs, revenues, habitat quality, wood flow, net present value, stream temperature levels, water quantity, forest structure quantity, and many others. Each of these types of quantitative metrics can be minimized or maximized, where appropriate. Once the decision has been made concerning an objective, a scheduling model would be developed to locate the best combination of activities that support minimizing or maximizing G.

A problem formulation ideally includes an *objective function* that is designed to evaluate the quality of a solution. The outcomes of the assignment of activities to decision variables add up to something, whether dollars, Euros, or habitat. An objective function would use these values to compare one potential solution to another, and to remember which of these is better. The objective function can be used as a measure of the *fitness* of a solution. *Decision variables* are needed to address forest harvest scheduling problems; they are the problem variables (e.g., whether to harvest a stand of trees) whose values are to be determined through the process employed in developing a feasible solution (Killen, 2021). Management decisions, represented by decision variables, need to be linked to outcomes to allow decision-makers the ability to compare alternative management scenarios (Iverson & Alston, 1986). During a forest harvest scheduling process, risk and uncertainty are often ignored or considered inherent to the data and assumptions employed. While maximization of an economic function is often based on the mean value of forest stands, other approaches can be used to account for risk, for example, by subtracting a risk premium from the value of each stand or by subtracting some other amount based on the variability inherent in a stand, in effect to represent risk aversion (Hahn et al., 2014). However, uncertainty may not be an issue of great concern when the data employed is of high quality and the time horizon is short (Nieuwenhuis &

Tiernan, 2005). Regardless of the recognition of risk or uncertainty, throughout the search for an optimal solution, the objective function value of each solution that is assessed provides the means to sort and rank the solutions.

3.2 Measurable Constraints and Accounting Rows

Nearly all practical problems have *constraints*, and if a solution is formed where they are violated, the solution is deemed infeasible to a problem (Michalewicz & Fogel, 2002). In forestry, constraints are often designed to control the assignment of activities to decision variables. For instance, if there were 20 potential ways to manage red pine stands in Wisconsin, a *resource constraint* may be designed to allow only one of them to be assigned to a management unit or stand. Or, if a red pine stand covered 50 acres (20.2 hectares) of land, a resource constraint may prevent harvesting more than 50 acres from this stand during the time horizon of the plan being developed. Without such constraints, a solution may suggest harvesting an infinite number of acres from this stand, even though that is not possible. In some harvest scheduling problems, resource constraints have been referred to as *absolute constraints*, where the amount of some resource is limited.

Accounting processes (e.g., the *accounting rows* employed in exact algorithms such as linear programming) are necessary to *add things up*. If it were necessary to know how much suitable wildlife habitat would be available during each time period of a forest harvest schedule, an accounting process (or a series of accounting rows) would be developed to provide this information. Similarly, one may be interested in understanding the amount of wood volume scheduled for harvest each year, or the standing (live) inventory of wood volume in a forest each year, or the amount of land area scheduled for each type of management action each year (final harvest, thinning, site preparation, etc.). Accounting processes or accounting rows do not act as constraints during a search process; however, the outcomes from employing them (e.g., the sum of the wood volume scheduled for harvest each year) can be constrained.

Policy constraints are used to control when and where management activities are allowed, and to control minimum and maximum amounts of outcomes produced by management activities. Policy constraints could be designed to use the outcomes from accounting processes to limit, or alternatively to require, a certain amount of wood volume to be scheduled during each time period of a time horizon. Wood flow constraints may be devised to ensure a consistent timber supply that is beneficial to local mills and local economies, which may in turn promote investment in processing facilities and associated infrastructure and prevent significant disruptions to the operation of various associated organizations in the local and regional economy (Stander et al., 2018). Wood flow constraints address two interesting aspects of a forest harvest scheduling problem: the idea of preferences (of the landowner or land manager) and the idea of forest production (Pukkala, 2002), where one is a human construct, and the other is a biological outcome. From the simple perspective of wood production, if harvests are planned in such a way that a balance is maintained between the growth of the younger forest stands and the harvest of the mature forest stands, then a forest can be considered managed under a sustained yield system. A desire by some landowners to control scheduled wood flow over time has been known for quite some time (Fernow, 1899;Schenck, 1907; Massachusetts Forest Service, 1909; Illinois Department of Conservation, State Division of Forestry, 1939; Committee on Interior and Insular Affairs, 1949). Efforts to estimate the maximum allowable harvest have also long been considered (Elliott & Mobley, 1949; Sandberg et al., 1949; Gross, 1950).

Example

In Europe, particularly Germany and Austria, concerns about the sustainable production of wood began to arise in the early- to mid-1700s (Morgenstern, 2007). In North America, concerns over a regular harvest of timber stemmed from a perceived timber famine that was supposed to occur in the early 1900s. Speculation proclaiming that a timber famine in the United States would be noticeable by the start of the twentieth century or thereabouts arose from many directions and was often based on little sound information (Koncen, 1881). This speculation inferred that states such as Maine, New York, and Pennsylvania would be economically and socially ruined (Brisbin, 1888). Around this time in the development of the United States, foresight for the sustainability of wood was lacking, and it was generally understood that harvests would continue without regard to the end of the supply, that harvest opportunities would depend only on market conditions and the capability of sawmills to process the wood (Roth, 1898). Timber supply estimates in California once suggested that only a 15-year supply of hardwood (deciduous) trees remained (Heath, 1913). Many eloquent and assertive statements were offered in general publications and governmental reports of the late nineteenth century to influence public opinion on the issue.

> *If any one doubts the danger of a timber famine in the United States at some future day, let him look at the destruction of trees in his own neighborhood. Where are the forests that sheltered your youth? Where are the big woods in which we hunted the red deer, the black and gray squirrel, and an occasional bear? Gone, gone, and all the game with them* (Brisbin, 1888).

> *... the time will come when supply will fall short of demand, is evident* (Leue, 1889).

A dire situation concerning a lack of wood never occurred as forest management organizations began to utilize methods that effectively stabilized forest resources (The New Hampshire Land Company, 1880). From this work, various methods to determine the maximum sustainable harvest level from a forest arose (or were adapted from the works of European foresters), based in part on desired future rotation ages, current and future standing growing stocks, growth rates, current and desired future age class distributions, desires

(continued)

of a wood-using facility, and others (Roth, 1914; Bettinger et al., 2017). The allowable cut or the annual sustained yield are two alternative metrics that can be used to reflect the maximization of a relatively consistent wood flow each year. The allowable cut is an old concept that has been applied to larger public land management organizations in North America for nearly a century. The concepts of an allowable cut and associated sustained yields have been viewed as a moral commitment or social obligation regarding the utilization, value gain, and growth of a forest resource (Wiebecke & Peters, 1984). At least from one perspective, these concepts were suggested as the ultimate objectives of a forest management plan (Illinois Central System, 1944).

Inventory constraints could also be designed to control how much standing wood volume exists in each time period. These types of policy constraints could be designed to ensure that the standing inventory in one time period (e.g., the first time period) is related to the standing inventory in another time period (e.g., the next time period). In addressing one aspect of forest sustainability, constraints can be employed to ensure that the standing inventory at the beginning of a time horizon might resemble the standing inventory at the end of a time horizon. Here, an ending inventory constraint might be used to prevent a harvest schedule from reducing the standing forest inventory to an unsustainable level and prevent liquidation of forest resources at the end of the time horizon (Garcia, 1990), particularly within problems that include a maximization objective. Methods employed can ensure a non-declining standing inventory across the time periods of the planning horizon (Stander et al., 2018), ensure that the ending inventory is greater than the average inventory level throughout the planning horizon, ensure that at no time the periodic inventory levels are below some proportion of the ending inventory, or perhaps ensure that a certain volume of harvestable growing stock is available after the end of the planning horizon (Government of Alberta, 2010).

Environmental constraints can further be designed to require, for example, a certain amount of suitable wildlife habitat during each time period of a forest plan, or to limit the number of negative outcomes (e.g., stream sediment produced) from the assignment of activities to units of land. Often these policy constraints are difficult to design given the complex nature of the functional relationships that exist between the outcomes (e.g., habitat, sediment) and the activities assigned to the landscape. Constraints such as these suggest that they *must be satisfied* to locate a feasible solution, and thus they are sometimes referred to as *hard constraints* in forest harvest scheduling problems (Michalewicz & Fogel, 2002).

There was a point in time in our recent past that large final harvests (clearcuts) were normal occurrences in certain rural parts of the world (Bersten et al., 1971). Although not well based in science, limiting the size of final harvests has been argued as a course of action that would be beneficial to aesthetic, soil, water, and wildlife concerns (National Forest Products Association, 1979). On the Jefferson National Forest in Virginia, a 40 acre (16.2 ha) maximum final harvest size is employed in certain areas (U.S. Department of Agriculture, Forest Service, 2004). At a lower governmental level in the United States, some final harvest size restrictions have been embedded in state laws. At the time this book was developed, no final harvest could exceed 250 acres (101.2 ha) in Maine (Maine Forest Service, 2014), and with certain exceptions, no final harvest could exceed 120 acres (48.6 ha) in Oregon (Oregon Secretary of State, 2020). In some parts of the world, final harvest sizes were even more restrictive, such as the 2.5–5 acres (1–2 ha) maximum size permitted in forests of the Czech Republic (Parliament of the Czech Republic, 1995).

The mathematical process used to arrange forest management decisions in a manner that best achieves an objective seeks to maximize or minimize some outcome subject to the condition that constraints have not been violated. To summarize this discussion, the key general concepts regarding a problem formulation for a model system are:

- *Time horizon*
- *Time period*
- *Objective function*
- *Accounting processes or rows*
- *Resource constraints*
- *Policy constraints*

4 Solving Forest Harvest Scheduling Assignment Problems

In one sense, the process used in the search for an optimal solution to a forest harvest scheduling problem can be viewed as *systematic* or *random*. A systematic search for the optimal solution to a problem might be considered a rational, meticulous evaluation of the possible options (Fig. 1.5). Complete enumeration, assessing every feasible solution to a problem, is the extreme case in this regard. However, some very rational, systematic processes use information gained through a search process to efficiently decide which further solutions to check (promising avenues that may lead to better solutions), and which further solutions to avoid (poor avenues with no chance of producing the optimal solution). A random search for the optimal solution to a problem (Fig. 1.6) may seem exciting, even irrational, particularly in the extreme case where every solution that is assessed has no relation to the previous set

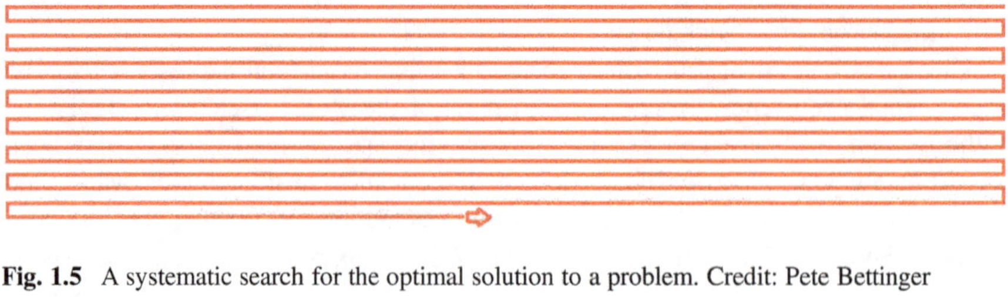

Fig. 1.5 A systematic search for the optimal solution to a problem. Credit: Pete Bettinger

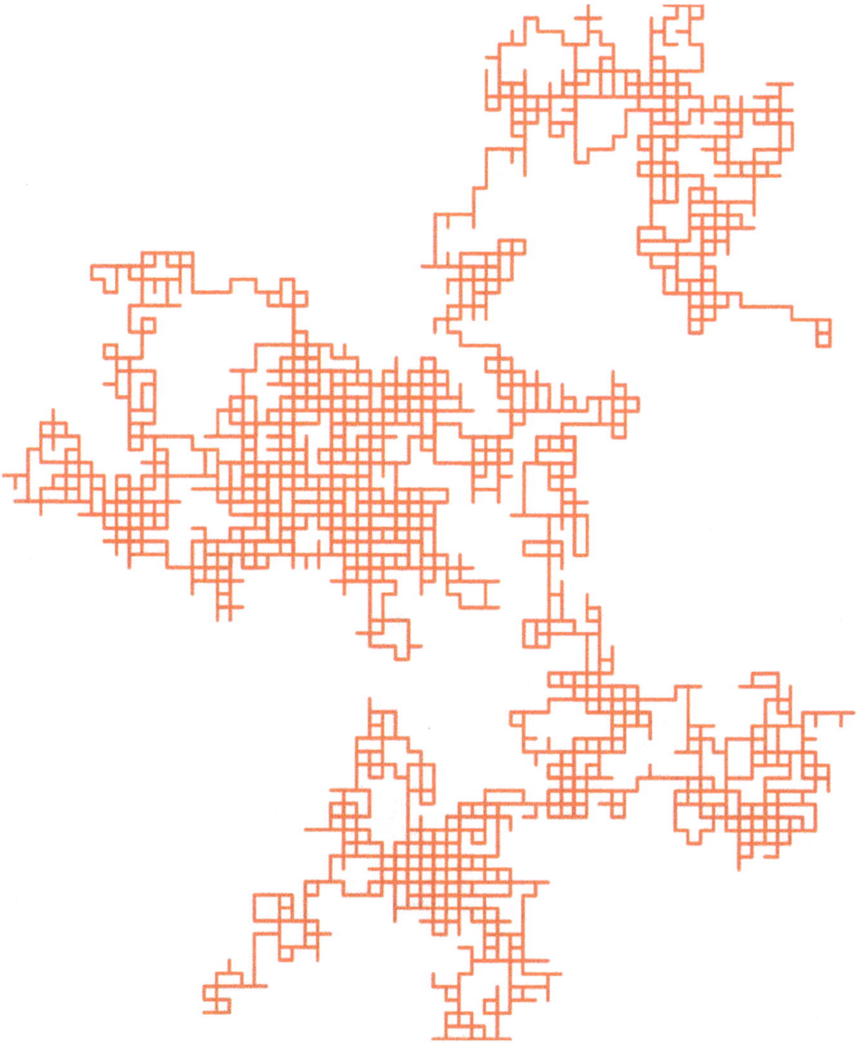

Fig. 1.6 A random search for the optimal solution to a problem. Credit: Oleg Alexandrov through Wikimedia Commons (https://commons.wikimedia.org/wiki/File:Random_walk_in2D_closeup.png)

of solutions that were assessed. Some random search processes can, however, be designed to take advantage of what has been learned during a search and use this information to quickly explore promising areas of the solution space.

With respect to search processes, *diversification* is the act of exploring parts of the solution space that have yet to be thoroughly explored. By contrast, *intensification* is the act of further exploring parts of the solution space where higher quality solutions have already been located. Similar to the act of searching for precious gems on a remote mountain in Colorado, diversification is prospecting (searching for good areas to mine) and intensification is mining (concentrating efforts in areas showing promise). Ideally, a search process would utilize both diversification and intensification tactics when exploring a solution space for the optimal solution to a problem. For some search processes, these tactics can be employed in a region of the solution space that is near the current solution to a problem, what is considered the local *neighborhood*. Neighboring solutions entail those that could be reached through a change to the status or assignment of one or more of the decision variables. In other cases, tactics can involve sampling a broader range of potential solutions. Without diversification, a search process may spend a great deal of time investigating potential solutions to a problem in a relatively poor area of a solution space. In this sense, the search process can become mesmerized by the prospect of good solutions and become trapped in an area around a local optimum. Alternatively, without intensification a search process may spend a great deal of time investigating potential solutions in random areas of the solution space. In this sense, the search process can become uninterested by the prospect of good solutions and spend too much time wandering aimlessly around the solution space.

As one may discover throughout this book, some of the forest harvest scheduling search processes described have behaviors that are indicative of both diversification and intensification. In fact, the behavior of search processes can shift from one type (diversification or intensification) to the other as conditions of the search change. Further, the timing (e.g., employing one type of search before the other) and the tenure (i.e., the amount of time one type of search is employed) can be important in adequately and efficiently conducting a search for an optimal solution. For example, it would not make sense to spend a lot of time intensively searching for the optimal solution around the first randomly located, feasible solution in the solution space; some diversification would seem worthwhile before intensifying a search. Further, once relatively high-quality solutions have been located in the solution space, it may not make much sense to spend the remaining computer processing time randomly searching other areas, without returning to the relatively high-quality places and spending more time searching for the optimal solution there.

4.1 The Size of the Solution Space

The solution space in many combinatorial problems can be enormous, and nearly infinite in content (Pohl, 1969). When decision variables can be assigned continuous real numeric values, theoretically, the number of different solutions to a problem approaches infinity. When decision variables need to be assigned integer values, the number of potential solutions in a solution space, from a forest management perspective, might be estimated as

$$\text{number of choices per stand}^{\text{stand}}$$

Stand 1	Stand 2	Stand 3	Stand 4	Stand 5	Comment	Stand 1	Stand 2	Stand 3	Stand 4	Stand 5	Comment
-	-	-	-	-	Harvest no stands	Cut	Cut	Cut	-	-	Harvest stands 1-3
Cut	-	-	-	-	Harvest stand 1	Cut	Cut	-	Cut	-	Harvest stands 1, 2, and 4
-	Cut	-	-	-	Harvest stand 2	Cut	Cut	-	-	Cut	Harvest stands 1, 2, and 5
-	-	Cut	-	-	Harvest stand 3	Cut	-	Cut	Cut	-	Harvest stands 1, 3, and 4
-	-	-	Cut	-	Harvest stand 4	Cut	-	Cut	-	Cut	Harvest stands 1, 3, and 5
-	-	-	-	Cut	Harvest stand 5	Cut	-	-	Cut	Cut	Harvest stands 1, 4, and 5
Cut	Cut	-	-	-	Harvest stands 1 and 2	-	Cut	Cut	Cut	-	Harvest stands 2, 3, and 4
Cut	-	Cut	-	-	Harvest stands 1 and 3	-	Cut	Cut	-	Cut	Harvest stands 2, 3, and 5
Cut	-	-	Cut	-	Harvest stands 1 and 4	-	Cut	-	Cut	Cut	Harvest stands 2, 4, and 5
Cut	-	-	-	Cut	Harvest stands 1 and 5	-	-	Cut	Cut	Cut	Harvest stands 3-5
-	Cut	Cut	-	-	Harvest stands 2 and 3	Cut	Cut	Cut	Cut	-	Harvest stands 1-4
-	Cut	-	Cut	-	Harvest stands 2 and 4	Cut	Cut	Cut	-	Cut	Harvest stands 1, 2, 3, and 5
-	Cut	-	-	Cut	Harvest stands 2 and 5	Cut	Cut	-	Cut	Cut	Harvest stands 1, 2, 4, and 5
-	-	Cut	Cut	-	Harvest stands 3 and 4	Cut	-	Cut	Cut	Cut	Harvest stands 1, 3, 4, and 5
-	-	Cut	-	Cut	Harvest stands 3 and 5	-	Cut	Cut	Cut	Cut	Harvest stands 2-5
-	-	-	Cut	Cut	Harvest stands 4 and 5	Cut	Cut	Cut	Cut	Cut	Harvest all stands

Fig. 1.7 Number of potential solutions involving 5 stands that have 2 choices each. Credit: Pete Bettinger

In the case of the Plantersville Tract mentioned earlier, with 16 potential choices for 81 different stands, the number of different forest harvest schedules is:

$$16^{81} = 3.41757925747346 \text{ E} + 97$$
$$16^{81} = 3.41757925747346 \times 10^{97}$$
$$16^{81} = 34{,}175{,}792{,}574{,}734{,}600{,}000{,}000{,}000{,}000{,}000{,}000{,}000{,}000{,}000{,}000,$$
$$000{,}000{,}000{,}000{,}000{,}000{,}000{,}000{,}000{,}000{,}000{,}000{,}000{,}000{,}000{,}000$$

Of course, many of these forest harvest schedules may be infeasible depending on the policy constraints imposed, but one does not know this without some prior information or some experience examining solutions in the solution space. To illustrate that estimates of the number of potential solutions are correct, imagine that there are only 5 stands, and that there are 2 choices for each (harvest, do not harvest). The number of potential solutions is

$$2^5 = 32$$

Figure 1.7 illustrates this set of potential solutions.

4.2 *The Manner in which Activities Are Assigned*

There are at least two distinct ways to represent a forest management problem mathematically and use the outcomes of it to provide a schedule of activities for field foresters to implement. The most straightforward of these was called *Model I* by Johnson and Scheurman (1977). In a Model I problem formulation, the identity of the management units (timber stands or strata) is retained throughout the time horizon. Johnson and Scheurman (1977) were more specific about this distinction in suggesting that an activity (a decision) refers to the entire set of management actions that apply to a land area over time. Versions of Model I allow multiple sequential activities to be assigned to a single land area, assuming the set is complementary over time, and assuming that the collection of activities maintains the integrity of the identity of the land area. In this case, the identity of the management unit would never be lost from the beginning of the time horizon to the end. In some sense, the spatial location of activities assigned (except in the fractional case) would be known (they are inside the timber stand). The design of the problem formulation in this manner seems trivial, but compared to other options, it has advantages (identity not lost) and disadvantages (perhaps requiring more detail).

A *Model II* problem formulation is conceptually different in the sense that land areas are not mathematically maintained as intact entities over time (Johnson & Scheurman, 1977), and thus multiple management units can be combined into a single strata (bin, collection of areas), and when this occurs, the spatial identity of the management units is lost. In some commercial harvest scheduling software, this grouping of similar management units (perhaps by their age or tree characteristics) occurs at the very beginning of the scheduling process. Activities (harvests) or management regimes might then be scheduled to portions of the total area contained in each stratum. While the grouping of management units into strata could occur prior to the initiation of a harvest scheduling process, or dynamically during the scheduling process, it is the latter case that defines a Model II problem according to Johnson and Scheurman (1977). During the scheduling process, management units (whole areas or parts thereof) that have final harvests scheduled in the same time period are grouped together *at that point in time* and managed as a different collective area from that point forward (Fig. 1.8), using a different decision variable.

A *Model III* problem formulation is further conceptually different and embodies a state trajectory perspective for structuring a model, where decision variables represent arcs within a decision tree, and these arcs contain only a beginning and ending state (Nobre et al., 2024). An interesting assessment and comprehensive taxonomic framework to classify works regarding the optimization of forest management

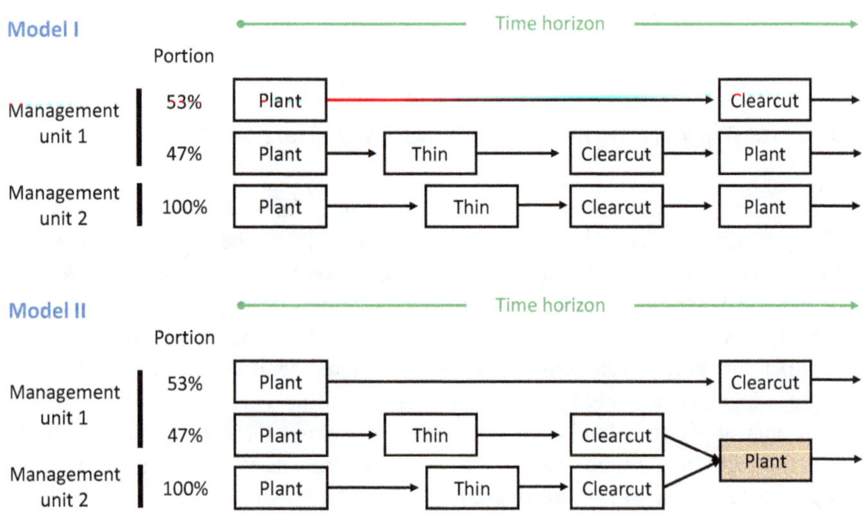

Fig. 1.8 Simple example of the difference between Model I and Model II. Credit: Pete Bettinger

activities based on model structure, methodology, and other factors can be found in Çağlayan et al. (2018).

4.3 Mathematical Programming Options (Exact or Algorithmic Methods)

Mathematical programming is the act of expressing a problem using mathematical notation, and subsequentially solving the problem using mathematical processes (Killen, 2021). These methods contain procedures that guarantee locating an optimal solution to problem (Romanycia & Pelletier, 1985), assuming the solution to the problem being solved is feasible. These approaches focus on allocating limited resources (time, money, etc.) to management activities or assigning decisions to units of management (machines, stands of trees) so that some overarching objective is maximized or minimized (i.e., optimized) and all of the associated constraints are satisfied (not violated). The classical approaches include linear programming, mixed integer programming (sometimes referred to 0–1 programming when the values allowed to be assigned to decision variables are binary), integer programming, linear goal programming, mixed-integer goal programming, and some forms of quadratic (non-linear programming) problems. These approaches rely on mathematics that ensure, when the processes have been completed, that the optimal solution to a problem has been located in a finite number of steps (Glover, 2022). In other words, if a management problem has been properly formulated, and if an exact algorithm has solved the problem, optimality is assured. Some characteristics of these methods include:

- Only one feasible solution is maintained as the *current solution* as an approach solves a problem.
- Only one feasible solution will be created from a *move* away from the current solution.[1]
- Only one solution is maintained in memory as the *best solution*, according to its objective function value.
- The moves are usually deterministic.

However, it has been understood for some time that problem formulations (models of real systems) may contain nonlinear relationships between proposed activities and desired outcomes, may refer to a solution space that is considered non-convex, and may have other characteristics (e.g., discrete or perhaps binary assignments) that suggest the problem may be difficult or impossible to solve using exact approaches (Dantzig, 1960). Significant efforts have been made regarding methods that intelligently reformulate complex non-linear problems into linear complements and

[1] This movement from one solution to another, through limited changes to decision variables, is said to have occurred within the nearby *neighborhood* of the current solution in the solution space.

remove redundancies so that these problems can be solved more easily, or in using continuous value ($0 - \infty$) variables that are assigned integer values without explicitly imposing these constraints (Glover, 1975; Adams et al., 2004). Mathematical programming (exact or algorithmic approaches) is widely applied in forest harvest scheduling. However, the rigidity of the nature of exact algorithms can be detrimental to an effective search of a solution space, and when a problem is combinatorically complex, some have argued that an effective search process must have some degree of freedom from rigid mathematical processes even though the identification of the global optimum solution to a problem might then be elusive (Glover, 1986).

Exact or algorithmic methods are the preferred approach for forest harvest scheduling when a management problem can be described sufficiently to fit the desired model structure. However, given the challenges noted earlier and environmental concerns that have arisen in the last 40 years which have prompted the inclusion of spatial constraints and integer decision variables in forest harvest scheduling efforts, some contemporary forest management problems have been proven difficult to accommodate with exact methods (Brodie & Sessions, 1991). As was shown earlier, the number of possible solutions to a complex problem may be so large that exhaustive search (complete enumeration) is impossible, and the problem may be so complicated that one might have to address relationships with non-linear functions or logic (if-then relationships). Further, there may also be cases where most of the local optima for a real world problem have poor value (Michalewicz & Fogel, 2002), or where discrete optimization problems are generally intractable (Johnson & Jacobson, 2002).

4.4 Heuristic Search Options (Heuristic Methods)

In instances where an exact or algorithmic method may present significant challenges for forest harvest scheduling efforts, a heuristic search process may be of value for developing high-quality forest plans (Glover, 2022). A heuristic search process is added to a basic problem-solving process (such as hill-climbing or random search) so that overall performance (however measured) might improve as the process explores paths toward an optimal solution (Romanycia & Pelletier, 1985). These search processes have been described as sets of general rules, independent of the problem type or domain, which can be used to solve problems (Franzin & Stützle, 2019). *Metaheuristics* have been described as higher-level processes that guide other heuristics (Glover, 2022). In this book, the two terms *heuristic* and *metaheuristic* are used interchangeably. For example, simulated annealing is described as a heuristic search process in Chap. 9, yet is it considered a point-based, s-metaheuristic. In general, a heuristic (or approximation) search process is used to obtain near-optimal solutions to difficult problems, employing a mathematical problem structure to locate feasible and near-optimal solutions, with mechanisms for avoiding becoming trapped in local optima. Heuristic search processes are designed to reduce the effort involved in problem-solving activity, yet they sacrifice the guarantee of an optimal solution provided by exact methods (Tonge, 1960). However, depending on the problem being addressed, some structure regarding the design of a heuristic search process is likely needed to efficiently sort through the potential feasible solutions in a solution space (Nurmela, 1993). The outcomes from a heuristic search process are therefore suboptimal, but often acceptable when the time or cost needed to locate the optimal solution to a problem are large (Hillier & Lieberman, 1980). A heuristic is considered effective if the objective function value of the feasible solutions produced is consistently near the optimal solution value (Bodin et al., 1981).

While operations research professionals often have a desire to characterize management problems in linear or mixed integer terms, pragmatic individuals who are *unencumbered by tradition* have attempted other computational methods, such as heuristic search processes, to solve difficult problems (Pohl, 1969). Much like Jennings' (1971) rules for success of corporate executives, a heuristic search process uses a strategy that, when it succeeds in locating a good solution, it likely will succeed in locating another nearby, and if, at any point, it fails to succeed in locating successively good solutions, it will likely break free and succeed in locating another good solution in another area of the solution space. Some heuristic search processes are inspired by the function of nature. One taxonomy of these types of search processes separates methods into those based on the behavior of individual organisms, those based on the behavior of swarms of organisms, and those based on functions related to laws of science and physical phenomena (Tzanetos et al., 2020). Another manner in which to describe heuristic search processes is to focus on the number of feasible solutions that are maintained during each iteration of the process. The idea of an *s-metaheuristic* is embodied in the ability of a search process to pivot from one feasible solution to another through limited changes to the assignment of activities to decision variables. The action of the pivot is often referred to as a *move* in the solution space. As an example, the two arrays noted in Fig. 1.9 represent feasible forest management plans and indicate the assignment of a management prescription to a forested stand. The second (to the right) was created through a minor change to the first. For a point-based, *s*-metaheuristic (e.g., threshold accepting, simulated annealing, tabu search) the *neighborhood* of moves includes the set of all potential feasible solutions that can be generated by transforming the current solution (Johnson & Jacobson, 2002). General characteristics of *point-based s*-metaheuristics (Talbi, 2009) include:

- Only one feasible solution is maintained as the *current solution* as an approach solves a problem.
- Only one feasible solution will be created from a *move* away from the current solution.
- Only one solution is maintained in memory as the *best solution*, according to its objective function value.
- A single potential change to the current solution is called a *1-opt move*.

Forest plan 1

Stand	Prescription assigned
1	23
2	12
3	5
4	17
5	24
6	23
7	8
8	11
9	7
10	13

Forest plan 2

Stand	Prescription assigned
1	23
2	12
3	5
4	17
5	3
6	23
7	8
8	11
9	7
10	13

Fig. 1.9 A change from one forest plan to another by way of a simple adjustment to the management prescription assigned to a stand. Credit: Pete Bettinger

- n-opt moves, which involve simultaneous changes to n decision variables, are possible.[2]
- The potential moves can be selected randomly or deterministically.
- The change in objective function value through the act of the potential move may determine whether the move will be accepted or rejected.
- The effect of the act of the potential move may create constraint violations, and thus determine whether the move will be accepted or rejected.[3]
- Thus, not all potential moves (changes to the current solution) will be acceptable.

A *p-metaheuristic* uses a set of feasible solutions that evolve during the search through limited to significant (perhaps even complete) changes to the assignment of activities to decision variables within each member (forest harvest schedule) of the set. This set of feasible solutions is sometimes referred to as a *population*. One example involves the selection of two different members of the population (two forest plans), splitting them, and recombining them into two new forest plans. Another example involves changing every member of the population based on its recent performance (with respect to changes in the objective function value), the

[2] In Chap. 8, Threshold Accepting, we delve further into the concept of n-opt moves.

[3] A caveat to this involves the use of strategic oscillation.

performance of all other members of the population, and changes observed in the value of the very best solution. These types of heuristic search processes are also called *population-based* metaheuristics (Talbi, 2009). Compared to *s*-metaheuristics, *p*-metaheuristics may allow more diversification during a search. General characteristics of population-based *p*-metaheuristics include:

- Many feasible solutions are maintained as *current solutions* in the population as an approach solves a problem.
- One or more feasible solutions will be created or modified from one or more of the current solutions in the population.[4]
- Only one solution is maintained in memory as the *best solution*, according to its objective function value.
- The potential creation or modification of a current solution can be conducted randomly or deterministically.
- The effect of creating or modifying a solution may result in constraint violations, and this may determine whether the created or modified solution will be accepted, rejected, or repaired.
- Thus, not all potential solutions that are created, modified, or repaired will be acceptable.

As with any of the heuristic search processes described in this book, generating just one solution to a problem is usually not good enough. Ideally, a very good heuristic would be used to generate a high number of solutions that are very close in quality to the global optimum. However, the performance of a heuristic can vary depending on the size and structure of each forest management problem. Therefore, until a sufficient number of tests have been conducted on a large group of problems, one may not be sure that one in two, or even one in 100, solutions generated by a heuristic are of sufficiently high quality.

5 Difficulty in Developing High Quality Forest Harvest Schedules

Although many forest harvest scheduling problems may seem conceptually simple, when emulating a real-world system, the number of variables and constraints can be very extensive. A problem is considered tractable (easy to deal with) if there exists an algorithm that can eventually solve it, and the computing time of the algorithm (or number of steps required) is bounded by a polynomial function of the size (n) of a problem (Karp, 1972). A tractable problem might be one where a set of linear equations can be devised for a forest harvest schedule, and the resulting problem formulation can be solved using an exact method (Karp, 1972). Theorists have

[4]In Chap. 11, Genetic Algorithm, we delve further into the concept of creating and modifying solutions of (or from) a population.

defined *P* as a complex class of tractable problems that can be solved in polynomial time (Tarjan, 1978). *Time* in this sense is represented by the number of computations conducted by a computing device, while the polynomial variable is related to the size of the data (Odifreddi, 2004). Unfortunately, many contemporary forest planning problems are intractable. Contemporary forest optimization problems which have wood flow and harvest adjacency constraints can require a great number of constraints (Bettinger, 2023) and can thus be difficult to formulate and solve exactly using mathematical programming approaches.

A problem is considered to belong to an *NP* (non-deterministic polynomial) class if its solution can be determined and verified by some non-deterministic polynomial time-bounded algorithms or a Turing machine (Hunt III, 1975). The *non-deterministic* aspect of this class implies that there is no particular rule employed to guess at the path to take to solve a problem (Tarjan, 1978). The amount of time required to arrive at the solution is considered relatively short, since it is again bounded by a polynomial function of the size of the problem. In theory, the optimal solution to problems in this class can be located using complete enumeration of all possible solutions, yet in practice, problems in this class may have an extremely large set of potentially feasible solutions (Wolsey, 1998). Unfortunately, difficult, or intractable problems may require computing time (or number of processing steps) that is an exponential function of the size (*n*) of a problem. In these cases, the effort required to solve a problem grows rapidly as the size of the problem increases. In many cases, models of real-world systems cannot be solved exactly in polynomial time (Rosenkrantz & Stearns, 2003) even today, with advances in computing systems that we enjoy. It may be impossible to find a polynomial time solution to a complex problem, or even show that one does not even exist (Odifreddi, 2004). Like many contemporary forest management problems, those that utilize integer and mixed integer programming problem formulations reside in this class (Karp, 1972).

Although difficult to prove, when a problem is considered *NP-hard*, there are no known polynomial-time algorithms that can be employed to solve it. Problems in this class are considered the most difficult problems of the *NP* class. *NP-hard* problems are often ones where the number of potential solutions increases rapidly as the problem size increases, due to the number of constraints that need to be simultaneously satisfied (Thalman et al., 1991). In the worst case, the effort expended to solve exactly (find the global optimum solution) *NP-hard* problems also increases exponentially with the size (number of decision variables, number of constraints) of the problem (Bodin et al., 1981) suggesting that it may be impractical to attempt to solve certain problems using exact algorithms (Bianchi et al., 2008). *NP-complete* problems are also those within the *NP* class where there are no available algorithms or methods that can locate a global optimum solution to a problem in polynomial time (i.e., *NP-hard*); however, solutions (local optima) can be located and verified in polynomial time (Velasco et al., 2022). *NP*-completeness is not necessarily a disadvantage for difficult problems, as approximation algorithms can be developed to produce solutions with satisfactory quality (closeness to the elusive optimal solution) in a satisfactory amount of time (Rosenkrantz & Stearns, 2003).

6 Summary

Quantitative harvest scheduling methods help a landowner or forestry organization achieve (at least mathematically) the desired condition of a future landscape, through the outcomes of the planned activities scheduled in pursuit of an objective and subject to a set of constraints (Killen, 2021). When forest managers follow a plan of action that recognizes objectives and constraints, they are ideally shaping the future landscape to the condition that is desired. Forest harvest scheduling processes need to represent a forested land base well and produce a plan of action that is efficient, using appropriate assumptions of potential management activities and the expected revenues and costs. However, even when the utmost care has been employed to develop a model within the limitations imposed by time, data, personnel, and organizational support, the application of harvest scheduling models can suffer from perceived biases favoring one resource value or another (Iverson & Alston, 1986).

Certainly, a trade-off is encountered between the amount and depth of technical detail contained in this book and the applicability of the ideas it provides to a broader audience in society. Forest harvest scheduling involves more than simply applying an established algorithm to a set of publicly available data. Cherry-picking and lifting heavily from the ideas expressed in Larson et al. (1990), the following statements concerning forest harvest scheduling likely hold true today, regardless of the type of landowner for which a forest harvest scheduling effort is being pursued. A forest harvest scheduling effort ...

- *Should faithfully characterize a forest's capabilities and limitations*
- *Should result in a clear guide for managing resources*
- *Should inform the allocation of scarce resources among competing needs*
- *Should inform the assignment of pertinent activities to amenable areas of land*
- *Should not be the exclusive domain of experts or their technical processes*
- *Should act to educate everyone involved*
- *Should limit the use of technical terms that can serve as a barrier to effective communication*
- *Might, however, encounter intractable or difficult problems*
- *Might, therefore, be an intense and contentious experience*

Take Home Message
Forest harvest scheduling is a specific endeavor of a broader set of work activities that may be required to develop a forest plan. The scheduling of forest management activities for different areas of land that contain trees (often called management units or stands) involves determining where to go, what to do, and when to conduct the activities. Methods range from very simple ad hoc processes to more complex mixed integer and heuristic search processes. In these efforts, the complexity of a real forest management problem is reduced to a mathematical model. Unfortunately, the resulting model may not completely describe the important issues (environmental, social, economic, legal) humans face when considering the management of land, as

some relationships between management activities and outcomes (primarily environmental and social) are difficult to address mathematically. However, a solution to the problem described in a model system can be of interest to forest and land managers and thus provide guidance for the management of a forest. This book describes many of the potential processes that can be employed for obtaining a forest harvest schedule. Ideally, the very best solution to a problem that is sought is one that consists of the very best arrangement of management activities over space and time.

Questions

1. With respect to forest harvest scheduling efforts, under what circumstances might the *Principle of Scale* be invalid?
2. How are forest-level and landscape-level planning processes different with respect to geography and land ownership?
3. How might an inventory constraint within a forest harvest scheduling problem be designed to address forest sustainability?
4. Aside from the need to directly supply a lumber mill with sawlogs, why are wood flow constraints important for forest harvest scheduling problems?
5. What might be required within the problem formulation of a forest harvest schedule to incorporate a policy constraint that controls the size of openings created by scheduled final harvests?
6. When integer decision variables are required in a forest harvest scheduling problem, which general search process might you select to solve the problem? Why?
7. What eventually may happen to the mathematical description of a forest stand when employing a Model II formulation in a forest harvest scheduling effort?
8. Considering the ability to navigate through a solution space, how might the behavior of linear programming differ from the behavior of an *s*-metaheuristic search process in solving a problem?

References

Adams, W. P., Forrester, R. J., & Glover, F. W. (2004). Comparisons and enhancement strategies for linearizing mixed 0-1 quadratic programs. *Discrete Optimization, 1*(2), 99–120.

Albers, D. J., Reid, C., & Dantzig, G. B. (1986). An interview with George B. Dantzig: The father of linear programming. *The College Mathematics Journal, 17*(4), 292–314.

Amery, C. F. (1875). *Notes on forestry*. Trubner & Co.

Bare, B. B., Briggs, D. G., Roise, J. P., & Schreuder, G. F. (1984). A survey of systems analysis models in forestry and the forest products industry. *European Journal of Operational Research, 18*(1), 1–18.

Bare, B. B., & Norman, E. L. (1969). *An evaluation of integer programming in forest production scheduling problems*. (Research Bulletin No. 847). Purdue University, Agricultural Experiment Station.

Bayfield County Forestry and Parks Committee. (2021). *Bayfield County forestry & parks comprehensive land use plan 2021–2035*. Bayfield County Forestry and Parks Committee.

Bellenguez, O., Brauner, N., & Tsoukiàs, A. (2023). Is there an ethical operational research practice? And what this implies for our research? *EURO Journal on Decision Processes, 11*, Article 100029.

Bersten, C. M., Litton, R. B., Jr., Lyon, L. J., Packer, P. E., Rees, P. M., & Schmautz, J. E. (1971). *Forest management in Wyoming: Timber harvest and the environment on the Teton, Bridger, Shoshone, & Bighorn National Forests*. U.S. Department of Agriculture, Forest Service, Wyoming Forest Study Team.

Bettinger, P. (1999). Distributing geographic information systems capabilities to field offices: Benefits and challenges. *Journal of Forestry, 97*(6), 22–26.

Bettinger, P. (2023). Modelling spatial connectivity of forest harvest areas: Exact and heuristic approaches. In C. Grueau, A. Rodrigues, & L. Ragia (Eds.), *Proceedings of the 9th International Conference on Geographical Information Systems Theory, Applications and Management (GISTAM 2023)* (pp. 136–143). SCITEPRESS—Science and Technology Publications, LDA.

Bettinger, P., Boston, K., Siry, J. P., & Grebner, D. L. (2017). *Forest management and planning* (2nd ed.). Academic Press.

Bettinger, P., Lennette, M., Johnson, K. N., & Spies, T. A. (2005). A hierarchical spatial framework for forest landscape planning. *Ecological Modelling, 182*, 25–48.

Bianchi, L., Dorigo, M., Gambardella, L. M., & Gutjahr, W. J. (2008). A survey on metaheuristics for stochastic combinatorial optimization. *Natural Computing, 8*(2), 239–287.

Bodin, L., Golden, B., Assad, A., & Ball, M. (1981). *The state of the art in the routing and scheduling of vehicles and crews (UMTA-SKED-URT41-81-1)*. U.S. Department of Transportation, Urban Mass Transportation Administration.

Bostedt, G., de Jong, J., Ekvall, H., Hof, A. R., Sjögren, J., & Zabel, A. (2021). An empirical model for forest landscape planning and its financial consequences for landowners. *Scandinavian Journal of Forest Research, 36*(7–8), 626–638.

Brisbin, J. S. (1888). *Trees and tree-planting*. Harper & Brothers.

Brodie, J. D., & Sessions, J. (1991). The evolution of analytic approaches to spatial harvest scheduling. In M. A. Buford (Comp.), *Proceedings of the 1991 Symposium on Systems Analysis in Forest Resources*. (General Technical Report SE-74), pp. 187–191. U.S. Department of Agriculture, Forest Service, Southeastern Forest Experiment Station.

Brown, J. (1894). *The Forester. A practical treatise on the planting and tending of forest trees and the general management of woodland estates, Volume II*. William Blackwood and Sons.

Burrows, G. B., Putnam, H. C., & Bruncken, E. (1898). *Report of the Forestry Commission of the State of Wisconsin*. Democrat Printing Company.

Çağlayan, İ., Yeşil, A., Çinar, D., & Cieszewski, C. (2018). Taxonomy for the optimization in forest management: A review and assessment. *Forestist, 68*(2), 122–135.

Carle, M. A., D'Amours, S., Azouzi, R., & Rönnqvist, M. (2020). A strategic forest management model for optimizing timber yield and carbon sequestration. *Forest Science, 67*(2), 205–218.

Chambers, C. J., Jr. (1977). *Sustainable harvest analysis, fiscal years 1978 and 1979*. (Harvest Regulations Series Report No. 7). Washington Department of Natural Resources.

Committee on Interior and Insular Affairs. (1949). *Sustained timber yield: Hearings before a subcommittee of the Committee on Interior and Insular Affairs, United States Senate, Eightieth Congress, Second Session*. U.S. Government Printing Office.

Curtis, F. H. (1962). Linear programming the management of a forest property. *Journal of Forestry, 60*(9), 611–616.

Cushen, W. E. (1955). Operations research and philosophy. *Science, 121*(3147), 584–586.

Dantzig, G. B. (1960). On the significance of solving linear programming problems with some integer variables. *Econometrica, 28*(1), 30–44.

Demirci, M., & Bettinger, P. (2015). Using mixed integer multi-objective goal programming for stand tending block designation: A case study from Turkey. *Forest Policy and Economics, 55*, 28–36.

Elliott, C. N., & Mobley, M. D. (1949). *Southern forestry*. Turner E. Smith & Co.

Epstein, R., Morales, R., Serón, J., & Weintraub, A. (1999). Use of OR systems in the Chilean forest industries. *Interfaces, 29*(1), 7–29.
Fadaei, A. H., Setayeshi, S., & Kia, S. (2009). An optimization method based on combination of cellular automata and simulated annealing for VVER-1000 NPP loading pattern. *Nuclear Engineering and Design, 239*(12), 2800–2808.
Fernow, B. E. (1899). *Forest policies and forest management in Germany and British India.* Government Printing Office.
Forest-Range Task Force. (1972). *The nation's range resources–A forest-range environmental study.* (Forest Resource Report No. 19). U.S. Department of Agriculture, Forest Service.
Franzin, A., & Stützle, T. (2019). Revisiting simulated annealing: A component-based analysis. *Computers & Operations Research, 104*, 191–206.
Freeman, O. E., Duguma, L. A., & Minang, P. A. (2015). Operationalizing the integrated landscape approach in practice. *Ecology and Society, 20*(1), Article 24.
French, S. (2023). Reflections on 50 years of MCDM: Issues and future research needs. *EURO Journal on Decision Processes, 11*, Article 100030.
Garcia, O. (1990). Linear programming and related approaches in forest planning. *New Zealand Journal of Forestry Science, 20*(3), 307–331.
Glover, F. (1975). Improved linear integer programming formulations of nonlinear integer problems. *Management Science, 22*(4), 455–460.
Glover, F. (1986). Future paths for integer programming and links to artificial intelligence. *Computers & Operations Research, 13*(5), 533–549.
Glover, F. W. (2022). Unforeseen consequences of "Tabu" choices–A retrospective. *INFORMS Journal on Computing, 34*(3), 1306–1308.
Government of Alberta. (2010). *C5 forest management plan 2006–2026*. Government of Alberta.
Gross, L. S. (1950). *Timber management plans on the national forests*. U.S. Department of Agriculture, Forest Service, Division of Timber Management.
Hahn, W. A., Hartl, F., Irland, L. C., Kohler, C., Moshammer, R., & Knoke, T. (2014). Financially optimized management planning under risk aversion results in even-flow sustained timber yield. *Forest Policy and Economics, 42*, 30–41.
Hann, D. W., & Bare, B. B. (1979). *Uneven-aged forest management: State of the art (or science?).* (General Technical Report INT-50). U.S. Department of Agriculture, Forest Service, Intermountain Forest and Range Experiment Station.
Heath, A. R. (1913). *The eucalyptus hardwood trees of California*. A.R. Heath.
Hillier, F. S., & Lieberman, G. J. (1980). *Introduction to operations research* (3rd ed.). Holden-Day, Inc.
Holmgren, P., & Thuresson, T. (1997). Applying objectively estimated and spatially continuous forest parameters in tactical planning to obtain dynamic treatment units. *Forest Science, 43*(3), 317–326.
Hunt, H. B., III. (1975). *On the complexity of finite, pushdown, and stack automata.* (MRC Technical Summary Report 1504). University of Wisconsin, Mathematics Research Center.
Illinois Central System. (1944). *Southern forestry. A study of the South's renewable natural resource*. Illinois Central System, Research and Development Bureau.
Illinois Department of Conservation, State Division of Forestry. (1939). *Elements of forestry with special reference to Illinois*. State of Illinois, Department of Conservation, State Division of Forestry.
Independent Multidisciplinary Science Team. (1999). *Recovery of wild salmonids in western Oregon forests: Oregon Forest Practices Act rules and the measures in the Oregon Plan for Salmon and Watersheds.* (Technical Report 1999-1). Oregon State University, Department of Forest Science, Independent Multidisciplinary Science Team.
Iverson, D. C., & Alston, R. C. (1986). *The genesis of FORPLAN: A historical and analytical review of Forest Service planning models.* (General Technical Report INT-214). U.-S. Department of Agriculture, Intermountain Research Station.
Jennings, E. E. (1971). *Routes to the executive suite*. McGraw-Hill Book Company.

Johnson, A. W., & Jacobson, S. H. (2002). On the convergence of generalized hill climbing algorithms. *Discrete Applied Mathematics, 119*(1–2), 37–57.

Johnson, K. N., & Scheurman, H. L. (1977). Techniques for prescribing optimal timber harvest and investment under different objectives—Discussion and synthesis. *Forest Science, Monograph, 18*.

Jones, J. G., Hyde, J. F. C., III, & Meacham, M. L. (1986). *Four analytical approaches for integrating land management and transportation planning on forest lands*. (Research Paper INT-361). U.S. Department of Agriculture, Forest Service, Intermountain Research Station.

Karp, R. M. (1972). Reducibility among combinatorial problems. In R. E. Miller & J. W. Thatcher (Eds.), *Complexity of computer computations* (pp. 85–103). Plenum Press.

Killen, J. (2021). *Mathematical programming methods for geographers and planners*. Routledge.

Koncen, M. (1881). *The people's protection party, with the following schemes, plans, and propositions, with additions as set forth in annex and appendix J*. Times Printing House.

Lacerda, T. H. S., de Jesus França, L. C., Lopez, I. L. E., Lacerda, S. L. S., Figueiredo, E. O., Barbosa, B. H. G., Silva, C. S. J. E., & Gomide, L. R. (2023). Multi-objective forest harvesting under sustainable and economic principles. *Journal of Forestry Research, 34*(5), 1379–1394.

Lämås, T., Sängstuvall, L., Öhman, K., Lundström, J., Årevall, J., Holmström, H., Nilsson, L., Nordström, E. M., Wikberg, P. E., Wikström, P., & Eggers, J. (2023). The multi-faceted Swedish Heureka forest decision support system: Context, functionality, design, and 10 years experiences of its use. *Frontiers in Forests and Global Change, 6*, Article 1163105.

Larson, G., Holden, A., Kapaldo, D., Leasure, J., Mason, J., Salwasser, H., Yonts-Shepard, S., & Shands, W. E. (1990). *Synthesis of the critique of land management planning, Volume 1 (FS-452)*. U.S. Department of Agriculture, Forest Service, Policy Analysis Staff.

Leue, A. (1889). *Fourth annual report of the Ohio-State Forestry Bureau to the Governor of the State of Ohio for the year 1888*. The Westbote Company.

Leuschner, W. A., Porter, J. R., Reynolds, M. R., & Burkhart, H. E. (1975). A linear programming model for multiple-use management. *Canadian Journal of Forest Research, 5*(3), 485–491.

Locher, A., Campa, H., III, Leefers, L., & Beyer, D. E., Jr. (2012). Understanding cumulative effects of aspen harvest on wildlife habitat and timber resources in northern Michigan. *Northern Journal of Applied Forestry, 29*(3), 113–127.

MacLeery, D. (1979). *Hearing before the Subcommittee on Parks, Recreation, and Renewable Resources of the Committee on Energy and Natural Resources, United States Senate, Ninety-Sixth Congress, First Session on the Oversight on Proposed National Forest Management Act regulations*. (Publication 96-29) (p. 47). U.S. Government Printing Office.

Maine Forest Service. (2014). *01-669, Maine Forest Service, Chapter 20, Forest regeneration and clearcutting standards*. Maine Department of Agriculture, Conservation, and Forestry, Maine Forest Service.

Marty, R. J. (1979). *Options for calculating allowable timber harvest*. (Research Report 371). Michigan State University, Agricultural Experiment Station.

Massachusetts Forest Service. (1909). *Forest working plan for land belonging to the City of Fall River on the North Watuppa Watershed*. Wright & Potter Printing Co.

Mendonça, N. P., Lopes, I. L. E., Gomes, V. S., Ferreira, M. A., Cruz, B. R., Silva, C. S. J. E., & Gomide, L. R. (2022). Silvicultural tasks scheduling optimization: A case study of functions and methods. *Revista Árvore, 45*, Article e4602.

Merganič, J., Merganičová, K., Výbošťok, J., Valent, P., & Bahýl, J. (2022). Impact of interest rates on forest management planning based on multi-criteria decision analysis. *Central European Forestry Journal, 68*(1), 23–35.

Michalewicz, Z., & Fogel, D. B. (2002). *How to solve it: Modern heuristics*. Springer-Verlag.

Morgenstern, E. K. (2007). The origin and early application of the principle of sustainable forest management. *The Forestry Chronicle, 83*(4), 485–489.

National Forest Products Association. (1979). *Hearing before the Subcommittee on Parks, Recreation, and Renewable Resources of the Committee on Energy and Natural Resources, United States Senate, Ninety-Sixth Congress, First Session on the Oversight on Proposed National*

Forest Management Act regulations. (Publication 96-29) (p. 49). U.S. Government Printing Office.

Nelson, J., & Brodie, J. D. (1990). Comparison of a random search algorithm and mixed integer programming for solving area-based forest plans. *Canadian Journal of Forest Research, 20*(7), 934–942.

Nieuwenhuis, M., & Tiernan, D. (2005). The impact of the introduction of sustainable forest management objectives on the optimisation of PC-based forest-level harvest schedules. *Forest Policy and Economics, 7*(4), 689–701.

Nobre, S. R., McDill, M. E., Rodriguez, L. C. E., & Diaz-Balteiro, L. (2024). Reframing forest harvest scheduling models for ecosystem services management. *Forests, 15*(12), 2236.

Nurmela, K. J. (1993). *Constructing combinatorial designs by local search*. (Series A: Research Reports No. 27). Helsinki University of Technology, Digital Systems Laboratory.

Odifreddi, P. (2004). *The mathematical century. The 30 greatest problems of the last 100 years*. Princeton University Press.

Oregon Secretary of State. (2020). *Department of Forestry, Chapter 629, Division 605, Planning forest operations*. Oregon Secretary of State.

Parliament of the Czech Republic. (1995). On forests and amendments to some acts (The Forest Act). In *Article 31, Regeneration and tending of forest stand, Parts 1a and 1b*. Parliament of the Czech Republic.

Pohl, I. (1969). *Bi-directional and heuristic search in path problems*. (SLAC Report No. 104). Stanford University, Stanford Linear Accelerator Center.

Pukkala, T. (2002). Introduction to multi-objective forest planning. In T. Pukkala (Ed.), *Multiobjective forest planning* (pp. 1–19). Kluwer Academic Publishers.

Pukkala, T., Kangas, J., Kniivilä, M., & Tiainen, A. M. (1997). Integrating forest-level and compartment-level indices of species diversity with numerical forest planning. *Silva Fennica, 31*(4), 417–429.

Reed, W. J., & Errico, D. (1986). Optimal harvest scheduling at the forest level in the presence of the risk of fire. *Canadian Journal of Forest Research, 16*(2), 266–278.

Romanycia, M. H., & Pelletier, F. J. (1985). What is a heuristic? *Computational Intelligence, 1*(1), 47–58.

Rosenkrantz, D. J., & Stearns, R. E. (2003). NP-complete problems. In A. Ralston, E. D. Reilly, & D. Hemmendinger (Eds.), *Encyclopedia of computer science* (pp. 1252–1256). John Wiley and Sons, Ltd.

Roth, F. (1898). *On the forestry conditions of northern Wisconsin*. (Bulletin No. 1, Economic Series No. 1). Wisconsin Geological and Natural History Survey.

Roth, F. (1914). *Forest regulation or the preparation and development of forest working plans*. (Volume I of Michigan Manual of Forestry). Filibert Roth.

Sandberg, L., Horn, A. G., & Guilkey, P. C. (1949). *The forest resource of Aitkin County*. Office of Iron Range Resources and Rehabilitation.

Sayer, J., Sunderland, T., Ghazoul, J., Pfund, J. L., Sheil, D., Meijaard, E., Venter, M., Boedhihartono, A. K., Day, M., Garcia, C., van Oosten, C., & Buck, L. E. (2013). Ten principles for a landscape approach to reconciling agriculture, conservation, and other competing land uses. *Proceedings of the National Academy of Sciences, 110*(21), 8349–8356.

Schenck, C. A. (1907). Forest management. In *Annual reports of the Forest, fish and game commissioner for 1904–1905-1906* (pp. 309–344). J.B. Lyon Company.

Selman, P. (1997). The role of forestry in meeting planning objectives. *Land Use Policy, 14*(1), 55–73.

Sen, T. K., & Chatterji, S. (2018). Land use planning: Basics and approaches. In G. Reddy & S. Singh (Eds.), *Geospatial technologies in land resources mapping, monitoring and management* (pp. 471–488). Springer.

Stander, H., Rasmussen, M., Greer, K., & Baribault, T. (2018). *Minnesota Department of Natural Resources sustainable timber harvest analysis phase 2 final report*. Mason, Bruce & Girard, Inc.

Steane, S. W. (1935). *A brief note on the principles of state forest policy*. Walter E. Shimmins, Government Printer.

Ström, I. A. (1822). Förslag till en förbättrad skogshushållning i Sverige, jemte utkast till dess systematiska verkställände (Proposal for an improved forest management in Sweden, together with a draft of its systematic implementation). Henry A. Nordström, Stockholm, Sweden.

Talbi, E. G. (2009). *Metaheuristics. From design to implementation*. John Wiley & Sons.

Tarjan, R. E. (1978). Complexity of combinatorial algorithms. *SIAM Review, 20*(3), 457–491.

Thalman, N., Sparn, T., Jaffres, L., Gablehouse, D., Judd, D., & Russell, C. (1991). AI techniques for a space application scheduling problem. In J. L. Rash (Ed.), *1991 Goddard Conference on Space Applications of Artificial Intelligence*. (NASA Conference Publication 3110) (pp. 83–93). National Aeronautics and Space Administration, Office of Management, Scientific and Technical Information Division.

The New Hampshire Land Company. (1880). *The American timber supply. Statement of the property in timberlands belonging to The New Hampshire Land Company; and its plan of organization*. The New Hampshire Land Company.

Thomas, P. R., & Salhi, S. (1998). A tabu search approach for the resource constrained project scheduling problem. *Journal of Heuristics, 4*(2), 123–139.

Tonge, F. M. (1960). *A heuristic program for assembly line balancing (P-1993)*. The RAND Corporation.

Tzanetos, A., Fister, I., Jr., & Dounias, G. (2020). A comprehensive database of nature-inspired algorithms. *Data in Brief, 31*, Article 105792.

U.S. Department of Agriculture, Forest Service. (2004). *Revised land and resource management plan, Jefferson National Forest*. (Management Bulletin R8-MB 115A). U.S. Department of Agriculture, Forest Service, Southern Region.

U.S. Department of Agriculture, Forest Service. (2014). *Revised land and resource management plan, George Washington National Forest*. (R8-MB 143A). U.S. Department of Agriculture, Forest Service, Region 8.

U.S. Department of Agriculture, Forest Service. (2016). *Tongass land and resource management plan. Final environmental impact statement. Plan amendment*. (R10-MB-769e,f). U.S. Department of Agriculture, Forest Service, Alaska Region, Tongass National Forest.

Ulvdal, P., Öhman, K., Eriksson, L. O., Wästerlund, D. S., & Lämås, T. (2023). Handling uncertainties in forest information: The hierarchical forest planning process and its use of information at large forest companies. *Forestry, 96*(1), 62–75.

Velasco, L., Guerrero, H., & Hospitaler, A. (2022). Can the global optimum of a combinatorial optimization problem be reliably estimated through extreme value theory? *Swarm and Evolutionary Computation, 75*, Article 101172.

Wiebecke, C., & Peters, W. (1984). Aspects of sustained yield history: Forest sustention as the principle of forestry—idea and reality. In H. K. Steen (Ed.), *History of sustained-yield forestry: A symposium* (pp. 176–182). Forest History Society, Inc.

Wieslander, A. E. (1922). *Management plan, Eastern Lassen working circle*. U.S. Forest Service, Lassen National Forest.

Wolsey, L. A. (1998). *Integer programming*. John Wiley & Sons, Inc.

Yoshimoto, A., & Brodie, J. D. (1994). Short- and long-term impacts of spatial restrictions on harvest scheduling with reference to riparian zone planning. *Canadian Journal of Forest Research, 24*(8), 1617–1628.

Zon, R. (1933). Land use planning by the states. *Journal of Farm Economics, 15*(2), 236–242.

Chapter 2
Linear Programming

1 Introduction

The origins of linear programming can be traced to works of European, North American, and Russian mathematicians who explored the ability of linear systems to address complex logistics and management problems of the mid-twentieth century. The work of Leonid Kantorovich (1960), which was presented in Russian in the late 1930s, provides perhaps the first effort to describe mathematical optimization to problems that may not have been previously perceived as mathematical in nature, including the assignment of tasks to machines used in the plywood industry (Koopmans, 1960). The linear programming process was formalized in 1947 by George Dantzig to describe a solution technique, *Programming within Linear Structure*, that he had developed to address organizational and logistical problems of a technical nature faced by the US military around the time of the Second World War (Wardle, 1965; O'Connor & Robertson, 2003; Odifreddi, 2004; Williams, 2005). Since then, and as a result of continual advances in computer technology and accessibility of the technology to the general public, linear programming has been successfully applied to a variety of societal problems. Over the last 60 years, the use of linear programming has gained widespread acceptance in forestry because forest planning problems are typically large (many decision variables) and complex (many interesting constraints), and these types of problems can often be formulated and solved in a computer system (Albers et al., 1986).

Some of the first works describing the potential of linear programming for assisting with the development of forest management plans became available in the early 1960s. For example, Coutu and Ellertsen (1960) described linear programming procedures to assess optimum alternatives for woodlands that were associated with small farms. Curtis (1962) later described how one might maximize scheduled wood production (timber harvest levels) or net present value subject to reforestation constraints. Large forest management organizations, such as the US Forest Service, began utilizing linear programming as a method for assisting in the development of

P. Bettinger, *Forest Harvest Scheduling*, Springer Texts in Business and Economics,
https://doi.org/10.1007/978-3-031-89432-9_2

management plans in the 1970s (US Department of Agriculture, Forest Service, 1978). A number of industrial companies, state forest management organizations, and Canadian Provinces continue to use variants of linear programming today to inform the development of forest plans.

2 Background

In a linear programming problem, one attempts to locate a vector x that best satisfies the following relationship:

$$A\,x = h$$

$$x > = 0$$

Here, matrix A would contain n rows (one row for each constraint or accounting row) and would contain $N > n$ columns. The two vectors, x and h, in this conceptual model have dimensions N and n, respectively. In solving the problem, the goal is to locate the best solution that is feasible within the structure of the constraints, which contains $(N - n)$ null elements, and where coefficients of the other elements are assigned some positive value (Householder, 1959). A *feasible solution* is one that is physically possible to implement, yet it may not necessarily be the optimal solution (Dantzig et al., 1954). The optimal solution is assumed to come from the set of feasible solutions.

When using linear programming, a model system (a forest, its production potential, accounting functions, and resource and policy constraints) is described by linear functions. One of these (the objective function) represents an idea that will either be maximized or minimized. The others represent ways of accumulating the outcomes of the planning process or ways of constraining the management activities that might be scheduled. The outcome is represented by an optimal combination of activities one might implement based on the value of the objective function. The main thrust of this process is that a model system can be described by sets of linear relations (Coutu & Ellertsen, 1960).

Three general types of outcomes can occur when using linear programming to address a planning problem designed for a model system (Leak, 1964). First, an optimal solution to a problem described with an objective function and a set of constraints may be located if the problem is appropriately designed. Second, when a problem is not appropriately designed, an infinite or unbounded result may occur. Some modern software products will notify users of this condition that occurs when the restrictions (often resource constraints) are not sufficient to create upper or lower bounds on the amounts of activities that are allowed. The third type of outcome is an infeasible solution, where the policy and resource constraints are inconsistent for describing a problem, and these inconsistencies prevent the location of any feasible solution. For example, one policy constraint may indicate that A + B $\geq$ 100, and two resource constraints may indicate that A $\leq$ 50 and B $\leq$ 40. Given these three requirements, there are no combinations of A and B that can satisfy the desires of the model.

As with all algorithms, there are some issues that require consideration before they can be employed in practice. For example, one assumption of linear programming involves the *divisibility* of the level of activity assigned to all of the decision variables. This assumption implies that all of the decision variables will be assigned a non-negative real value that reflects the magnitude of the decision. Although logically some decisions should be represented by a discrete value (e.g., harvest the entire stand during a single time period), with linear programming, the decision variables are assigned continuous real values from zero to infinity (unless otherwise constrained).

Example

In the development of a 15-year forest harvest schedule for a southern US forest, there might be a number of management regime options available to each stand, and the options can be independent and not synergistic. Ideally, only one of the options should be selected for each stand. However, a solution to a problem designed using linear programming might suggest that 47% of a stand (or some other fractional value) will be managed in one manner (management regime A) and 53% of a stand will be managed in another (management regime B). Interestingly, the location of where the 47% or the 53% is located in a stand of trees is unknown. As another example, if the decisions available were to simply harvest (clearcut) a stand of trees, or not harvest a stand of trees, linear programming may assign a fraction of the whole (e.g., 33%) to the harvest decision during one period of time, and the remainder (67%) to a harvest decision during another period of time. Often, these instances of the divisibility of a logical decision occur as a result of wood flow constraints.

A second assumption of linear programming involves the *certainty* with which we know the coefficients assigned to each decision variable. Although for some time the potential application of stochastic linear programming (Näslund, 1965) has been acknowledged, we assume in this text that the certainty assumption prevails. With this assumption, the coefficients associated with each decision variable in the objective function, the constraints, and the accounting rows are static (e.g., they do not change as the problem is being solved). In addition, the right-hand side values of each constraint are also known and static. With respect to forestry the tree volumes, prices, costs, habitat, carbon, and other outcomes of management that are important are known before the linear programming process is applied and do not change as the process attempts to solve a problem. Some forms of stochastic linear programming also make these assumptions, yet they differ from standard linear programming in that a set of solutions are generated after randomly changing certain coefficients, in order to observe how the objective function value might change when uncertainty is acknowledged (Näslund, 1965).

A third assumption of linear programming involves the *proportionality* of the value of decision variables contained in the objective function. Imagine that the development of one mile of road will cost $25,000 and one acre (0.404 ha) of site preparation activities will cost $300. In an objective function that minimizes a budget and that includes road construction decisions (25,000 × road mile developed) and site preparation decisions (300 × acres site prepared), each mile of road, no matter how many miles of road are constructed, will cost the same amount ($25,000) and each acre of site prepared land, no matter how many are treated, will cost the same amount ($300). The contribution of each decision variable to the objective function value of the solution is therefore proportional to the costs (or benefits in other cases) assumed and does not change as different decisions are made, nor as the magnitude of the same decision changes.

The fourth assumption of linear programming involves the *additivity* of each contribution from each decision variable to the objective function. In an objective function, such as $30X + 20Y + 43Z$, the contribution of any of the decisions (X, Y, or Z) is known (30, 20, or 43), and any one of these contributions is independent of the other two decisions that might be made. An optimal solution that suggests the production of 1000 units of X generates provides a value of 30 units to the solution every time, no matter how many Y or Z might be produced.

3 Process

Linear programming allows one to develop efficient alternatives to complex problems in forestry and natural resource management. Through the adjustment of decision variable coefficients and constraint amounts, alternative forest plans can be assessed rather quickly. In general, the process might be described in this manner:

- Locate an initial basic feasible solution, often a null (no assignments made) set.
- While there are variables that are "not in the solution" which might contribute to a better solution:
  - Assess from the set of variables whether any one of them can improve the objective function value.

    If none exists, terminate the process and report the optimal solution to the problem.

  - Select the variable that can result in the greatest increase in the objective function value.
  - Increase the value of this variable to the maximum contribution possible.
  - Adjust the value of all other decision variables, to maintain feasibility. One variable (either a real-decision variable or a slack variable) will be forced out of the solution.

3.1 Objective Function, Constraints, and Accounting Rows

An optimization problem inherently should contain an expression that will be optimized. This expression is presented mathematically, with the goal of locating the best combination and number of items (*decision variables*) to produce that will either maximize its value or minimize its value. In forestry, one might seek to maximize wood production, or net present value, or wildlife habitat, as long as the expression that defines these is related to the decisions that need to be made (what to do, if anything, to each stand of trees within the time horizon of a forest plan). On the other hand, one might seek to minimize costs related to the management of a forest or to minimize some environmental impact such as sediment produced from logging operations or the use of roads. The *objective function* thus contains the *decision variables* (or choice variables) and their corresponding value, or *objective coefficients* (Leuschner et al., 1975; Sierksma, 1996).

Constraints limit the number of items that can be produced. In a general sense, *resource constraints* limit the production of items based on how many might be available. For example, a resource might be related to the budget available to conduct forestry-related work. A resource constraint may then limit the amount spent so that the plan does not suggest spending more than what might be contained in a budget. Another resource constraint might be related to land areas. For example, constraints might be imposed to limit how much of a stand of trees can be harvested (i.e., no more than 100%). Similarly, constraints might be imposed on a problem to limit how many times (rather than percentage) a stand of trees can be entered (e.g., at most once). *Policy constraints* are related to organizational decisions, laws, regulations, and other issues that need to be recognized in a plan of action. For example, a policy constraint may be devised to ensure that the scheduled wood flow (harvest amount) is less than or equal to the growth of a forest. Or policy constraints can be devised to limit the size (amount or physical geographic size) of aggregated management activities.

The left-hand side of a constraint (the part to the left of the equal sign or the inequality sign) is represented by a linear function containing some (or all) of the decision variables and their corresponding coefficients (value within each constraint). The right-hand side of a constraint is represented by a value that represents threshold that should (minimum amount) or should not (maximum amount) be exceeded (thus the constraint) or an exact value to be met. The right-hand side of a constraint cannot contain a variable in a linear programming model. The objective function coefficients, constraint-related coefficients, and right-hand side values of constraints are assumed to be continuous real numbers (Sierksma, 1996) rather than integers, although they may appear to be the latter.

Accounting rows in a linear programming model are designed simply to add things up. Accounting rows themselves do not act to constrain the solution to a management problem. Perhaps the total amount of land assigned to each type of management activity in each year of a forest harvest schedule might be of interest to a landowner or forest manager. Similarly, the total amount of scheduled wood harvest might be of interest. Accounting rows are designed to accumulate values, and once the total amounts are determined, these accumulated values can be constrained using policy or resource constraints.

Example
Consider the following problem.

```
Maximize 12 A + 23 B + 17 C
Subject to
A ≤ 35
B ≤ 55
C ≤ 40
A + B ≤ 70
A + C ≤ 66
B + C ≤ 80
77 A + 33 B + 49 C ≤ 6020
A + B + C - TOTAL = 0
```

The objective function is represented by the first equation. Here, it suggests that a function (*12 A + 23 B + 17 C*) will be maximized as items *A*, *B*, and *C* are produced. In a maximization problem like this, the ideal solution would be to produce an infinite number (∞) of these items, resulting in an unbounded objective function value. However, as is indicated in the problem, some amounts of *A*, *B*, or *C* might be produced, and the first three constraints limit the amounts of *A*, *B*, and *C* that can be produced. These are resource constraints. The next four equations are policy constraints. The final equation is an accounting row that accumulates the amounts of *A*, *B*, and *C* produced. An accounting row such as this allows the further development of policy constraints that utilize the new variable that is introduced (here, *TOTAL*) to represent the sum of the values assigned to other decision variables. For example, an equation that could be added to the problem is

```
TOTAL ≤ 100
```

For a maximization problem, the ideal solution would be to produce the greatest amount of the items represented in the objective function, therefore the set of resource constraints are devised to limit the amounts produced to those that are reasonable or biologically possible. For example, if a forestry problem was devised to suggest that Stand 1 could produce 5000 units of wood in time period 1 if it were clearcut (assigned a final harvest treatment), or that Stand 1 could produce 5100 units of wood in time period 2 if it were clearcut, a resource constraint would be designed to prevent both of these actions from occurring if it were unreasonable to assume a stand of trees could be scheduled for a final harvest in two consecutive time periods.

3.2 Detached Coefficient Matrix

The assumptions of divisibility, certainty, proportionality, and additivity allow the linear programming process to operate very efficiently when a management problem is described. The problem, once it is formalized as a set of linear equations, is transformed into a detached coefficient matrix. The *basis matrix* is simply a structure containing the coefficients related to the decision variables (columns) and constraints and accounting functions (rows), and each iteration of the Simplex Method (described below) produces a new basis that differs from the previous using simple recursion relations that transform the values (Dantzig, 1953). The idea is to manipulate the columns of the detached coefficient matrix so that after a finite number of iterations (movements from vertex to vertex of the solution space), the optimal solution is located (Sierksma, 1996). Some characteristics of a detached coefficient matrix include:

1. The first row is reserved for the objective function.
2. Each of the subsequent rows represents the constraints and the accounting rows.
3. One row is needed for each constraint or accounting row.
4. The total number of rows is the sum of the number of constraints and the number of accounting rows, plus 1 (the objective function).
5. One column is needed for each decision variable.
6. One column is needed for the right-hand side value of each constraint or accounting row.
7. For every constraint that includes an inequality (e.g., ≤, ≥), a slack variable is created.
8. One column is needed for each slack variable.
9. The total number of columns is the sum of the number of decision variables and the number of slack variables, plus 1 (the right-hand side values).

Example

Consider the following problem again.

```
Maximize 12 A + 23 B + 17 C
Subject to
A ≤ 35
B ≤ 55
C ≤ 40
A + B ≤ 70
A + C ≤ 66
B + C ≤ 80
```

(continued)

```
77 A + 33 B + 49 C ≤ 6020
A + B + C - TOTAL = 0
```

For visual purposes, where variables within equations and the right-hand side values are stacked vertically, a modified arrangement of the problem is presented in Fig. 2.1.

The problem needs to be converted to canonical form, and thus the constraints that have an inequality sign need to be converted to constraints that contain an equality sign. To accomplish this, a separate *slack variable* is introduced within each constraint. The slack variables can be assigned continuous real numbers that range in value from 0 to infinity (∞). For example,

```
1 A + 1 B ≤ 70
```

is technically the same as

```
1 A + 1 B + Slack1 = 70
```

In this case, the sum of both *A* and *B* has to equal 70 or less to satisfy the constraint, as was suggested in the original constraint equation. If the sum of both *A* and *B* variables is over 70, the constraint is violated because the slack variable cannot be negative. For example, if the sum of both *A* and *B* is 73, the slack variable would need to be −3, which is not possible since the range of valid values is 0 to ∞. If the sum of both *A* and *B* variables is less than 70, there exists some slack in the constraint. For example, perhaps *A* and *B* are both 30, in this case *Slack1* has to be 10 units to satisfy the constraint. When the slack value related to a constraint is 0, the constraint is considered *binding*. In this case, the constraint is not binding because $A + B = 60$, which is 10 less than the right-hand side (thus there are 10 units of slack related to the constraint before it becomes binding).

Had the constraint included a greater than or equal to operator (≥), the slack variable would have a negative (subtraction) sign in front, indicating that

```
1 A + 1 B ≥ 70
```

would have been translated to

```
1 A + 1 B - Slack1 = 70
```

In this case, the sum of both *A* and *B* has to equal 70 or greater to satisfy the constraint, as was suggested in the original constraint equation. If the sum of

(continued)

both *A* and *B* variables is less than 70, the constraint is violated because the slack value would have to be negative. If the sum of both *A* and *B* variables is greater than 70, there exists some slack in the constraint. For example, perhaps *A* and *B* are both 45, in this case *Slack1* would have to be 20 units to satisfy the constraint. As one can imagine, after inserting the slack variables into each inequality constraint, the problem becomes physically larger (Fig. 2.2).

After removing the alphanumeric variable names, the summation signs (+), and the equal signs, we arrive at the detached coefficient matrix (Fig. 2.3), where the coefficients are now separated from their variables (detached). In addition, all of the nine characteristics noted above hold particularly the need for one column to represent each of the variables, one row to represent each of the constraints and slack variables, and the top row to represent the objective function.

Fig. 2.1 A modified arrangement of the simple linear programming problem. Credit: Pete Bettinger

12 A	+ 23 B	+ 17 C			
1 A				<=	35
	1 B			<=	55
		1 C		<=	40
1 A	+ 1 B			<=	70
1 A		+ 1 C		<=	66
	1 B	+ 1 C		<=	80
77 A	+ 33 B	+ 49 C		<=	6020
1 A	+ 1 B	+ 1 C	- 1 TOTAL	=	0

12 A	+ 23 B	+ 17 C										
1 A				+ 1 Slack1							=	35
	1 B				+ 1 Slack2						=	55
		1 C				+ 1 Slack3					=	40
1 A	+ 1 B						+ 1 Slack4				=	70
1 A		+ 1 C						+ 1 Slack5			=	66
	1 B	+ 1 C							+ 1 Slack6		=	80
77 A	+ 33 B	+ 49 C								+ 1 Slack7	=	6020
1 A	+ 1 B	+ 1 C	- 1 TOTAL								=	0

Fig. 2.2 An expanded view of the simple linear programming problem after slack variables have been included. Credit: Pete Bettinger

12	23	17									
1				1							35
	1				1						55
		1				1					40
1	1						1				70
1		1						1			66
	1	1							1		80
77	33	49								1	6020
1	1	1	-1								0

Fig. 2.3 The detached coefficient matrix for the simple linear programming model. Credit: Pete Bettinger

3.3 The Simplex Method

The Simplex Method (or *simplex procedure* or *simplex process* or *simplex algorithm*) was once characterized as a *finite iterative procedure* (or *method*) for solving, within a finite number of iterations, problems that contain a set of non-negative decision variables and linear constraints (Dantzig, 1953; Dantzig et al., 1954, 1956).

> *The term "simplex" evolved from an early geometrical version in which (like in game theory) the variables were non-negative and summed to unity. In that formulation a class of "solutions" was considered which lay in a simplex* (Dantzig et al., 1954).

The Simplex Method method is but one technique for solving a linear programming problem, yet it has been lauded for its ability to address the computational issues that are associated with management problems formulated as linear systems (Bare & Norman, 1969). The Simplex Method is a process (Fig. 2.4) for algebraically transforming a detached coefficient matrix and solving a linear programming problem such that the final answer represents the optimal solution (most efficient) to the problem. The Simplex Method consists of a set of mathematical operations that conceptually move from one extreme point (corner) of the solution space to another, as long as the resulting solution has an objective function value equal to or greater than the previous extreme point (corner). These extreme points (corners) of the solution space are generally considered *basic feasible solutions*. A few associated concepts that are more deeply described in operations research texts include the following:

- *Basic variable*: A variable that has a non-zero value assignment.
- *Basis*: The set of basic variables.
- *Basic solution*: A solution to a problem that satisfies the technical constraints (is feasible) but not necessarily the non-negativity constraints. The number of variables in the final solution are less than or equal to the number of technical constraints. At each corner point of the solution space, the number of non-zero variables (those "in the solution" will at most equal the number of constraints).

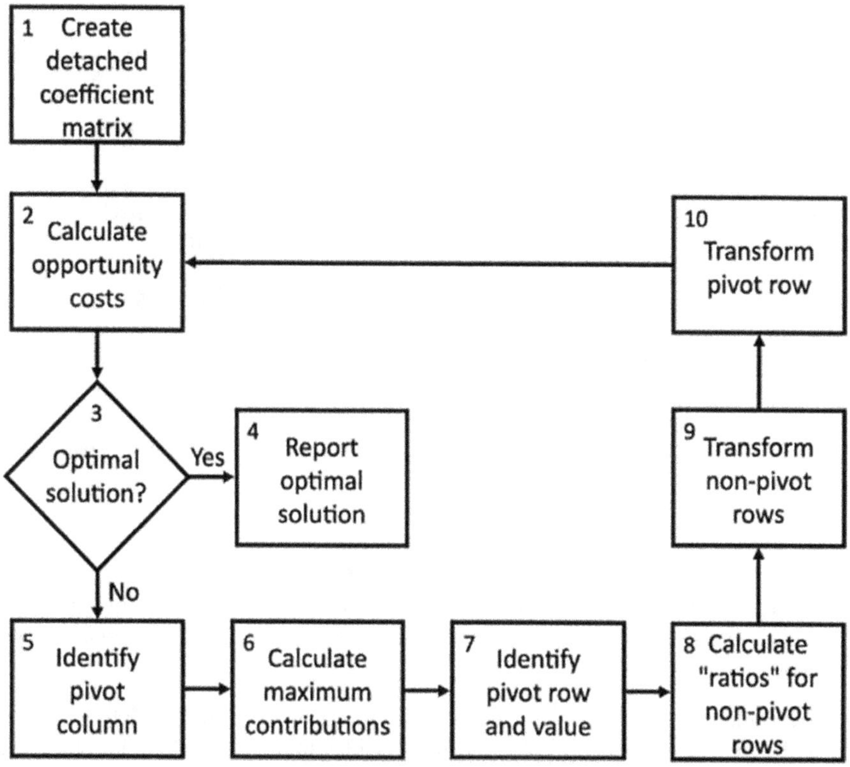

Fig. 2.4 One characterization of the process involved when employing the Simplex Method. Credit: Pete Bettinger

- *Basic feasible solution*: A basic solution that also satisfies the non-negativity constraints.

If an optimal solution to a problem exists, the mathematical procedures within the Simplex Method will eventually arrive at the extreme point (corner) most distant from the origin, with respect to the objective function value, for a maximization problem. The steps within the Simplex Method introduce one variable into the basis and remove another variable (slack or otherwise) from the basis. Employing the method provides an optimal basis (set of actions to follow) as well as proof of optimality (Avis, 2015). In this part of the chapter, an application of the Simplex Method to a maximization problem will be described.

Using the previously described example, once a detached coefficient matrix has been developed, for the purposes of this explanation of the process, some additional information is presented (Fig. 2.5). An indication across the top of the matrix is provided concerning the column related to each variable, and two additional columns on the right are provided. The first column (second from far right) indicates the variable that is present in the current feasible solution. The second column (far right)

	A	B	C	TOTAL	Slack1	Slack2	Slack3	Slack4	Slack5	Slack6	Slack7	Quantity	Variables in the solution	Variable value to the solution
Row 1	12	23	17											
Row 2	1				1							35	Slack1	0
Row 3		1				1						55	Slack2	0
Row 4			1				1					40	Slack3	0
Row 5	1	1						1				70	Slack4	0
Row 6	1		1						1			66	Slack5	0
Row 7		1	1							1		80	Slack6	0
Row 8	77	33	49								1	6020	Slack7	0
Row 9	1	1	1	-1								0	TOTAL	0

Fig. 2.5 The detached coefficient matrix enhanced with additional annotation. Credit: Pete Bettinger

indicates how much that variable is worth to the solution. In other words, this column simply reflects the objective function value of each of the variables that are present in the current solution. Annotation on the left side of the matrix, listing the row numbers, was also provided for the discussion that follows. A few important issues should be noted at this point:

- The initial point in the *n*-dimensional polytope where the Simplex Method begins is the *null set*. In other words, for the variables that are noted in the objective function (*A*, *B*, and *C*), each is initially assigned the value of zero (0).
- The original right-hand side values of each constraint and accounting row are now listed as *quantities*. These quantities will change as the detached coefficient matrix is transformed using the Simplex Method.
- For each row in the detached coefficient matrix, with the exception of the objective function row (row 1 of the detached coefficient matrix), the variables in the initial solution are the slack variables and the *TOTAL* variable.
- The amount assigned to variables in the initial solution (as well as future solutions) is reflected by the quantity (third column from the right) that is displayed to the left of the variable name in the solution. For example, *Slack1* = 35 and *TOTAL* = 0 in the initial solution. These results should be obvious, since if *A*, *B*, and *C* are all 0 in the initial solution, then for the first constraint (row 2 of the matrix) to be true, *Slack1* must be 35. Further, if *A*, *B*, and *C* are all 0, then the *TOTAL* must also be zero.
- The slack variables and the *TOTAL* variable all have no value in the solution, as suggested in the far-right column of Fig. 2.5. In other words, these variables are not present in the objective function, therefore when the objective function is calculated, they contribute nothing to it.

Step 2 of the Simplex Method (Fig. 2.4) involves determining the *opportunity costs* associated with each decision variable. Since each decision variable is represented by a column in the detached coefficient matrix, an opportunity cost is calculated for each column:

$$\text{Opportunity cost}_{\text{Column X}} = \text{Objective function value of variable in } \textit{Column X} - \sum_{y=2}^{n} \left((\textit{Column X Row Y} \text{ value}) \times \left(\textit{Row Y} \text{ ``variable value to the solution''} \right) \right)$$

The objective function values of the variables can be located in Row 1 of the detached coefficient matrix. Extending the opportunity cost function to represent the problem we have been following thus far,

$$\begin{aligned}\text{Opportunity cost}_{\text{A}} = {} & \text{Objective function value of variable} A \\ & - \left((\textit{Column A Row } 2 \text{ value}) \times \left(\textit{Row } 2 \text{ ``variable value to the solution''} \right) \right) \\ & - \left( (\textit{Column A Row } 3 \text{ value}) \times \left( \textit{Row } 3 \text{ ``variable value to the solution''} \right) \right) \\ & - \left((\textit{Column A Row } 4 \text{ value}) \times \left(\textit{Row } 4 \text{ ``variable value to the solution''} \right) \right) \\ & - \left( (\textit{Column A Row } 5 \text{ value}) \times \left( \textit{Row } 5 \text{ ``variable value to the solution''} \right) \right) \\ & - \left((\textit{Column A Row } 6 \text{ value}) \times \left(\textit{Row } 6 \text{ ``variable value to the solution''} \right) \right) \\ & - \left( (\textit{Column A Row } 7 \text{ value}) \times \left( \textit{Row } 7 \text{ ``variable value to the solution''} \right) \right) \\ & - \left((\textit{Column A Row } 8 \text{ value}) \times \left(\textit{Row } 8 \text{ ``variable value to the solution''} \right) \right) \\ & - \left( (\textit{Column A Row } 9 \text{ value}) \times \left( \textit{Row } 9 \text{ ``variable value to the solution''} \right) \right)\end{aligned}$$

or more specifically,

$$\begin{aligned}\text{Opportunity cost}_{\text{A}} = {} & 12 \\ & - ((1) \times (0)) \\ & - ((0) \times (0)) \\ & - ((0) \times (0)) \\ & - ((1) \times (0)) \\ & - ((1) \times (0)) \\ & - ((0) \times (0)) \\ & - ((77) \times (0)) \\ & - ((1) \times (0))\end{aligned}$$

where ultimately,

$$\text{Opportunity cost}_{\text{A}} = 12$$

If one conducts the mathematics for the other columns in the detached coefficient matrix, it will be determined that

$$\text{Opportunity cost}_{B} = 23$$
$$\text{Opportunity cost}_{C} = 17$$
$$\text{Opportunity cost}_{TOTAL} = 0$$
$$\text{Opportunity cost}_{Slack1} = 0$$
$$\text{Opportunity cost}_{Slack2} = 0$$
$$\text{Opportunity cost}_{Slack3} = 0$$
$$\text{Opportunity cost}_{Slack4} = 0$$
$$\text{Opportunity cost}_{Slack5} = 0$$
$$\text{Opportunity cost}_{Slack6} = 0$$
$$\text{Opportunity cost}_{Slack7} = 0$$
$$\text{Opportunity cost}_{Slack8} = 0$$

When the opportunity costs are all either zero or a negative value, the optimal solution to a problem has been located. This is the decision rule associated with the question in Step 3 of the Simplex Method process (Fig. 2.4). Once this condition is true (all opportunity costs are zero or negative), the Simplex Method terminates and the optimal solution, as represented by the *variables in the solution* and their associated *quantities*, is reported. However, if at least one opportunity cost that arises from Step 2 of the Simplex Method process is a positive value, it is understood that the optimal solution to a problem has not yet been located. In these cases, the basic feasible (current) solution can be improved by inserting an activity currently absent from it into the solution (Wardle, 1965). If the optimal solution to the problem has not yet been located, the decision variable with the greatest opportunity cost is selected to *enter the solution* in this respect. The column related to this decision variable then becomes the *pivot column* (Step 5 in Fig. 2.4). In the case we have been following, decision variable B would be selected to enter the solution since it had the highest opportunity cost (23 units).

The motivation for choosing the decision variable to enter the solution is based on a desire to improve the objective function value (Sierksma, 1996). With respect to the decision variable entering the solution, the maximum amount (or *maximum contribution*) of this variable that can enter the solution is then determined, using

- Each constraint row (not accounting row) in the detached coefficient matrix
- Each constraint row where the variable B is represented (i.e., rows 3, 5, 7, and 8 in this example)

$$\text{Maximum contribution}_{ROW} = (\text{Current quantity}_{ROW} / \text{Pivot column value}_{ROW})$$

For the problem we have been following thus far,

$$\text{Maximum contribution}_{\text{ROW 3}} = (55/1) = 55$$
$$\text{Maximum contribution}_{\text{ROW 5}} = (70/1) = 70$$
$$\text{Maximum contribution}_{\text{ROW 7}} = (80/1) = 80$$
$$\text{Maximum contribution}_{\text{ROW 8}} = (6020/33) = 182.42$$

The maximum contributions indicate how many units, at most, of the selected decision variable can be inserted into the solution. The row that indicates the most limiting amount of the selected decision variable is delegated as the *pivot row* (Step 7 in Fig. 2.4). In this case, Row 3 indicates that, at most, only 55 units of decision variable B can enter the solution. Therefore, this is the most constraining of the four options, and this row is therefore deemed the pivot row for this iteration of the Simplex Method. All other rows of the detached coefficient matrix, with the exception of Row 1 where the objective function coefficient values reside, are considered the *non-pivot rows*. Shortly, the values in both the pivot row and the non-pivot rows will be transformed. The values in the row representing the objective function will not be transformed.

The intersection of the pivot column and the pivot row, for the example illustrated here, is the value (*key number*) which will be used as a divisor to transform all of the values in the pivot row. At the moment, the key number is trivial (1).

To transform the values in the non-pivot rows, *ratios* for each of the non-pivot rows need to be developed (Step 8 in Fig. 2.4). A ratio, for each row in the detached coefficient matrix (except the objective function row), is determined by dividing whatever value is located in the pivot column on a non-pivot row by the key number. Each non-pivot row may have a different ratio. For the first iteration of the Simplex Method, when applied to the problem noted above, these again are trivial computations.

$$\text{Ratio}_{\text{ROW 2}} = (\text{value of Row 2 in the pivot column/key number}) = (0/1) = 0$$
$$\text{Ratio}_{\text{ROW 4}} = (\text{value of Row 4 in the pivot column/key number}) = (0/1) = 0$$
$$\text{Ratio}_{\text{ROW 5}} = (\text{value of Row 5 in the pivot column/key number}) = (1/1) = 1$$
$$\text{Ratio}_{\text{ROW 6}} = (\text{value of Row 6 in the pivot column/key number}) = (0/1) = 0$$
$$\text{Ratio}_{\text{ROW 7}} = (\text{value of Row 7 in the pivot column/key number}) = (1/1) = 1$$
$$\text{Ratio}_{\text{ROW 8}} = (\text{value of Row 8 in the pivot column/key number}) = (33/1) = 33$$
$$\text{Ratio}_{\text{ROW 9}} = (\text{value of Row 9 in the pivot column/key number}) = (1/1) = 1$$

Once the ratios are known for each row, all of the values in the non-pivot rows are transformed (Step 9 in Fig. 2.4) in this manner:

$$\text{New non}-\text{pivot row value}_{\text{ROW X COLUMN Y}}=\text{Old non}-\text{pivot row value}_{\text{ROW X COLUMN Y}} \\ -(\text{Pivot row value}_{\text{COLUMN Y}}\times\text{Ratio}_{\text{ROW X}})$$

As an example, for Row 8 of the detached coefficient matrix, the following new matrix values in this non-pivot row are determined in this manner:

$$\text{New non}-\text{pivot row value}_{\text{ROW 8 COLUMN 1}}=\text{Old non}-\text{pivot row value}_{\text{ROW 8 COLUMN 1}} \\ -(\text{Pivot row value}_{\text{COLUMN 1}}\times\text{Ratio}_{\text{ROW 8}})=(77-(0\times33))=77$$
$$\text{New non}-\text{pivot row value}_{\text{ROW 8 COLUMN 2}}=\text{Old non}-\text{pivot row value}_{\text{ROW 8 COLUMN 2}} \\ -(\text{Pivot row value}_{\text{COLUMN 2}}\times\text{Ratio}_{\text{ROW 8}})=(33-(1\times33))=0$$
$$\text{New non}-\text{pivot row value}_{\text{ROW 8 COLUMN 3}}=\text{Old non}-\text{pivot row value}_{\text{ROW 8 COLUMN 3}} \\ -(\text{Pivot row value}_{\text{COLUMN 3}}\times\text{Ratio}_{\text{ROW 8}})=(49-(0\times33))=49$$
$$\text{New non}-\text{pivot row value}_{\text{ROW 8 COLUMN 4}}=\text{Old non}-\text{pivot row value}_{\text{ROW 8 COLUMN 4}} \\ -(\text{Pivot row value}_{\text{COLUMN 4}}\times\text{Ratio}_{\text{ROW 8}})=(0-(0\times33))=0$$
$$\text{New non}-\text{pivot row value}_{\text{ROW 8 COLUMN 5}}=\text{Old non}-\text{pivot row value}_{\text{ROW 8 COLUMN 5}} \\ -(\text{Pivot row value}_{\text{COLUMN 5}}\times\text{Ratio}_{\text{ROW 8}})=(0-(0\times33))=0$$
$$\text{New non}-\text{pivot row value}_{\text{ROW 8 COLUMN 6}}=\text{Old non}-\text{pivot row value}_{\text{ROW 8 COLUMN 6}} \\ -(\text{Pivot row value}_{\text{COLUMN 6}}\times\text{Ratio}_{\text{ROW 8}})=(0-(1\times33))=-33$$
$$\text{New non}-\text{pivot row value}_{\text{ROW 8 COLUMN 7}}=\text{Old non}-\text{pivot row value}_{\text{ROW 8 COLUMN 7}} \\ -(\text{Pivot row value}_{\text{COLUMN 7}}\times\text{Ratio}_{\text{ROW 8}})=(0-(0\times33))=0$$
$$\text{New non}-\text{pivot row value}_{\text{ROW 8 COLUMN 8}}=\text{Old non}-\text{pivot row value}_{\text{ROW 8 COLUMN 8}} \\ -(\text{Pivot row value}_{\text{COLUMN 8}}\times\text{Ratio}_{\text{ROW 8}})=(0-(0\times33))=0$$
$$\text{New non}-\text{pivot row value}_{\text{ROW 8 COLUMN 9}}=\text{Old non}-\text{pivot row value}_{\text{ROW 8 COLUMN 9}} \\ -(\text{Pivot row value}_{\text{COLUMN 9}}\times\text{Ratio}_{\text{ROW 8}})=(0-(0\times33))=0$$
$$\text{New non}-\text{pivot row value}_{\text{ROW 8 COLUMN 10}}=\text{Old non}-\text{pivot row value}_{\text{ROW 8 COLUMN 10}} \\ -(\text{Pivot row value}_{\text{COLUMN 10}}\times\text{Ratio}_{\text{ROW 8}})=(0-(0\times33))=0$$
$$\text{New non}-\text{pivot row value}_{\text{ROW 8 COLUMN 11}}=\text{Old non}-\text{pivot row value}_{\text{ROW 8 COLUMN 11}} \\ -(\text{Pivot row value}_{\text{COLUMN 11}}\times\text{Ratio}_{\text{ROW 8}})=(1-(0\times33))=1$$
$$\text{New non}-\text{pivot row value}_{\text{ROW 8 COLUMN 12}}=\text{Old non}-\text{pivot row value}_{\text{ROW 8 COLUMN 12}} \\ -(\text{Pivot row value}_{\text{COLUMN 12}}\times\text{Ratio}_{\text{ROW 8}})=(6020-(55\times3))=4205$$

The pivot row values are then transformed (Step 10 in Figure 2.4) by dividing all previous pivot row values by the key number.

$$\text{New pivot row value}_{\text{COLUMN Y}}=(\text{Old pivot row value}_{\text{COLUMN Y}}/\text{key number})$$

This is a trivial exercise at this point (iteration 1 of the Simplex Method) for this example, since the key number was identified to have a value of 1.

After the pivot row and the non-pivot row values have been transformed, we have completed a single iteration of the Simplex Method. Conceptually, we have moved around the polytope representing the *n*-dimensional solution space, from one corner to another neighboring corner. The newly transformed detached coefficient matrix (Fig. 2.6) indicates that product B is in the solution, with a quantity of 55 units, the maximum amount, according to constraint 2, of product B that could be produced. This variable has a value in the objective function; therefore, the far-right column was updated to reflect that each unit of product B produced is worth 23 toward the objective function value. The quantities of *Slack4* (now 15) and *Slack6* (now 25) have also changed since product B is now being produced. The quantity of *TOTAL* is also now −55, which seems odd at first glance. However, when this value is placed

	A	B	C	TOTAL	Slack1	Slack2	Slack3	Slack4	Slack5	Slack6	Slack7	Quantity	Variables in the solution	Variable value to the solution
Row 1	12	23	17											
Row 2	1				1							35	Slack1	0
Row 3		1				1						55	B	23
Row 4			1				1					40	Slack3	0
Row 5	1					- 1		1				15	Slack4	0
Row 6	1		1						1			66	Slack5	0
Row 7			1			- 1				1		25	Slack6	0
Row 8	77		49			- 33					1	4205	Slack7	0
Row 9	1		1	- 1		- 1						- 55	TOTAL	0

Fig. 2.6 Results of the first transformation of the detached coefficient matrix. Credit: Pete Bettinger

into constraint 7 ($A + B + C - TOTAL = 0$), the real total is 55 units (the same as the amount of product B currently being produced).

The newly transformed detached coefficient matrix is then used to determine a new set of opportunity costs (Step 2 of Fig. 2.4). The process computes again the opportunity costs to determine whether the optimal solution has been located, otherwise the marginal contributions, pivot row, etc., are re-identified and the detached coefficient matrix is once again transformed.

$$\text{Opportunity cost}_{\text{A}} = 12$$
$$\text{Opportunity cost}_{\text{B}} = 0$$
$$\text{Opportunity cost}_{\text{C}} = 17$$
$$\text{Opportunity cost}_{\text{TOTAL}} = 0$$
$$\text{Opportunity cost}_{\text{Slack1}} = 0$$
$$\text{Opportunity cost}_{\text{Slack2}} = -23$$
$$\text{Opportunity cost}_{\text{Slack3}} = 0$$
$$\text{Opportunity cost}_{\text{Slack4}} = 0$$
$$\text{Opportunity cost}_{\text{Slack5}} = 0$$
$$\text{Opportunity cost}_{\text{Slack6}} = 0$$
$$\text{Opportunity cost}_{\text{Slack7}} = 0$$

From this analysis, it is concluded that (1) the optimal solution has not yet been located (some opportunity costs are positive), (2) product C will enter the solution (highest opportunity cost), and (3) column 3 is the pivot column. For the problem we have been following thus far, each constraint row where product C is located is assessed for the maximum contribution (amount) of product C can be inserted into the current solution to the problem.

$$\text{Maximum contribution}_{\text{ROW 4}} = (40/1) = 40$$
$$\text{Maximum contribution}_{\text{ROW 6}} = (66/1) = 66$$
$$\text{Maximum contribution}_{\text{ROW 7}} = (25/1) = 25$$
$$\text{Maximum contribution}_{\text{ROW 8}} = (4205/49) = 85.82$$

	A	B	C	TOTAL	Slack1	Slack2	Slack3	Slack4	Slack5	Slack6	Slack7	Quantity	Variables in the solution	Variable value to the solution
Row 1	12	23	17											
Row 2	1				1							35	Slack1	0
Row 3		1				1						55	B	23
Row 4						1	1			- 1		15	Slack3	0
Row 5	1					- 1		1				15	Slack4	0
Row 6	1					1			1	- 1		41	Slack5	0
Row 7			1			- 1				1		25	C	17
Row 8	77					16				- 49	1	2980	Slack7	0
Row 9	1			- 1						- 1		- 80	TOTAL	0

Fig. 2.7 Results of the second transformation of the detached coefficient matrix. Credit: Pete Bettinger

Row 7 of the detached coefficient matrix is identified as the pivot row, since at most, 25 units of product C can be inserted into the solution. The key number at the intersection of the pivot column and the pivot row is 1. The ratios for the non-pivot rows are:

$$\text{Ratio}_{\text{ROW } 2} = 0$$
$$\text{Ratio}_{\text{ROW } 3} = 0$$
$$\text{Ratio}_{\text{ROW } 4} = 1$$
$$\text{Ratio}_{\text{ROW } 5} = 0$$
$$\text{Ratio}_{\text{ROW } 6} = 1$$
$$\text{Ratio}_{\text{ROW } 8} = 49$$
$$\text{Ratio}_{\text{ROW } 9} = 1$$

Using this information, the detached coefficient matrix is once again transformed (Fig. 2.7). The opportunity costs are again determined.

$$\text{Opportunity cost}_{\text{A}} = 12$$
$$\text{Opportunity cost}_{\text{B}} = 0$$
$$\text{Opportunity cost}_{\text{C}} = 0$$
$$\text{Opportunity cost}_{\text{TOTAL}} = 0$$
$$\text{Opportunity cost}_{\text{Slack1}} = 0$$
$$\text{Opportunity cost}_{\text{Slack2}} = -6$$
$$\text{Opportunity cost}_{\text{Slack3}} = 0$$
$$\text{Opportunity cost}_{\text{Slack4}} = 0$$
$$\text{Opportunity cost}_{\text{Slack5}} = 0$$
$$\text{Opportunity cost}_{\text{Slack6}} = -17$$
$$\text{Opportunity cost}_{\text{Slack7}} = 0$$

As with the previous iteration of the Simplex Method, from this analysis, it is concluded that the optimal solution has not yet been located (some opportunity costs are positive). Here, product A will be entering into the solution (highest opportunity cost), and column 1 is the pivot column. For the problem we have been following thus far, each constraint row where product A is located is assessed for the maximum contribution (amount) of product A can be inserted into the current solution to the problem.

$$\text{Maximum contribution}_{\text{ROW } 2} = (35/1) = 35$$
$$\text{Maximum contribution}_{\text{ROW } 5} = (15/1) = 15$$
$$\text{Maximum contribution}_{\text{ROW } 6} = (41/1) = 41$$
$$\text{Maximum contribution}_{\text{ROW } 8} = (2980/77) = 38.7$$

Row 5 of the detached coefficient matrix is identified as the pivot row, since at most, 15 units of product A can be inserted into the solution. The key number at the intersection of the pivot column and the pivot row is 1. The ratios for the non-pivot rows are:

$$\text{Ratio}_{\text{ROW } 2} = 1$$
$$\text{Ratio}_{\text{ROW } 3} = 0$$
$$\text{Ratio}_{\text{ROW } 4} = 0$$
$$\text{Ratio}_{\text{ROW } 6} = 1$$
$$\text{Ratio}_{\text{ROW } 7} = 0$$
$$\text{Ratio}_{\text{ROW } 8} = 77$$
$$\text{Ratio}_{\text{ROW } 9} = 1$$

Using this information, the detached coefficient matrix is once again transformed (Fig. 2.8). The opportunity costs are once again determined.

	A	B	C	TOTAL	Slack1	Slack2	Slack3	Slack4	Slack5	Slack6	Slack7	Quantity	Variables in the solution	Variable value to the solution
Row 1	12	23	17											
Row 2					1	1		- 1				20	Slack1	0
Row 3		1				1						55	B	23
Row 4						1	1			- 1		15	Slack3	0
Row 5	1					- 1		1				15	A	12
Row 6						2		- 1	1	- 1		26	Slack5	0
Row 7			1			- 1				1		25	C	17
Row 8						93		- 77		- 49	1	1825	Slack7	0
Row 9				- 1		1		- 1		- 1		- 95	TOTAL	0

Fig. 2.8 Results of the third transformation of the detached coefficient matrix. Credit: Pete Bettinger

$$\text{Opportunity cost}_{\text{A}} = 0$$
$$\text{Opportunity cost}_{\text{B}} = 0$$
$$\text{Opportunity cost}_{\text{C}} = 0$$
$$\text{Opportunity cost}_{\text{TOTAL}} = 0$$
$$\text{Opportunity cost}_{\text{Slack1}} = 0$$
$$\text{Opportunity cost}_{\text{Slack2}} = 6$$
$$\text{Opportunity cost}_{\text{Slack3}} = 0$$
$$\text{Opportunity cost}_{\text{Slack4}} = -12$$
$$\text{Opportunity cost}_{\text{Slack5}} = 0$$
$$\text{Opportunity cost}_{\text{Slack6}} = -17$$
$$\text{Opportunity cost}_{\text{Slack7}} = 0$$

As with the previous iterations of the Simplex Method, from this analysis, it is concluded that the optimal solution has not yet been located. Interestingly, slack 2 has a positive opportunity cost, therefore column 6 is the pivot column. The maximum contribution (amount) of slack 2 that can be inserted into the current solution to the problem at this point in time, using only the positive values of the maximum contributions, is 13. Therefore, Row 6 will be the pivot row.

$$\text{Maximum contribution}_{\text{ROW 2}} = (20/1) = 20$$
$$\text{Maximum contribution}_{\text{ROW 3}} = (55/1) = 55$$
$$\text{Maximum contribution}_{\text{ROW 4}} = (15/1) = 15$$
$$\text{Maximum contribution}_{\text{ROW 5}} = (15/-1) = -15$$
$$\text{Maximum contribution}_{\text{ROW 6}} = (26/2) = 13$$
$$\text{Maximum contribution}_{\text{ROW 7}} = (25/-1) = -25$$
$$\text{Maximum contribution}_{\text{ROW 8}} = (1825/93) = 19.62$$

Row 6 of the detached coefficient matrix is identified as the pivot row, since at most, 13 units of slack 2 can be further inserted into the solution. The key number at the intersection of the pivot column and the pivot row is 2. The ratios for the non-pivot rows are:

$$\text{Ratio}_{\text{ROW 2}} = 0.5$$
$$\text{Ratio}_{\text{ROW 3}} = 0.5$$
$$\text{Ratio}_{\text{ROW 4}} = 0.5$$
$$\text{Ratio}_{\text{ROW 5}} = -0.5$$
$$\text{Ratio}_{\text{ROW 7}} = -0.5$$
$$\text{Ratio}_{\text{ROW 8}} = 46.5$$
$$\text{Ratio}_{\text{ROW 9}} = 0.5$$

Using this information, the detached coefficient matrix is once again transformed (Fig. 2.9). For a fifth time, the opportunity costs are again determined.

	A	B	C	TOTAL	Slack1	Slack2	Slack3	Slack4	Slack5	Slack6	Slack7	Quantity	Variables in the solution	Variable value to the solution
Row 1	12	23	17											
Row 2					1			- 1	- 1	1		7	Slack1	0
Row 3		1						1	- 1	1		42	B	23
Row 4							1	1	- 1	- 1		2	Slack3	0
Row 5	1							1	1	- 1		28	A	12
Row 6						1		- 1	1	- 1		13	Slack2	0
Row 7			1					- 1	1	1		38	C	17
Row 8								- 31	-47	- 3	1	616	Slack7	0
Row 9				- 1				- 1	-1	- 1		- 108	TOTAL	0

Fig. 2.9 Results of the fourth transformation of the detached coefficient matrix. Credit: Pete Bettinger

$$\text{Opportunity cost}_{\text{A}} = 0$$
$$\text{Opportunity cost}_{\text{B}} = 0$$
$$\text{Opportunity cost}_{\text{C}} = 0$$
$$\text{Opportunity cost}_{\text{TOTAL}} = 0$$
$$\text{Opportunity cost}_{\text{Slack1}} = 0$$
$$\text{Opportunity cost}_{\text{Slack2}} = 0$$
$$\text{Opportunity cost}_{\text{Slack3}} = 0$$
$$\text{Opportunity cost}_{\text{Slack4}} = -9$$
$$\text{Opportunity cost}_{\text{Slack5}} = -3$$
$$\text{Opportunity cost}_{\text{Slack6}} = -14$$
$$\text{Opportunity cost}_{\text{Slack7}} = 0$$

From an inspection of the opportunity costs, one can conclude that the optimal solution to the problem has been determined. Once the optimal solution has been located, the current value of the decision variables is reported to describe the components of the optimal solution.

$$\text{Objective function value} = 1948$$
$$\text{Product A} = 28 \text{ units}$$
$$\text{Product B} = 42 \text{ units}$$
$$\text{Product C} = 38 \text{ units}$$
$$\text{TOTAL} = 108 \text{ units}$$

The optimal solution also indicates that there is some slack in the right-hand side of a few of the constraints.

Slack 1 = 7 units (constraint#1 of the problem formulation where $A \leq 35$)
Slack 2 = 13 units (constraint#2 of the problem formulation where $B \leq 55$)
Slack 3 = 2 units (constraint#3 of the problem formulation where $C \leq 40$)
Slack 7 = 616 units (constraint#7 of the problem formulation where $77A + 33B + 49C \leq 6020$)

In sum, the Simplex Method is initialized somewhere in the solution space (likely with a null set of activities selected) and using pivot rules, the process deterministically locates an efficient path through the solution space, moving from corner to corner through the n-dimensional polytope (Avis, 2015). The solution to the problem lies on the boundary of the solution space, and the search process focuses on those feasible solutions located at the intersection of a subset of the planes formed by the constraints (Householder, 1959).

However, if one of the decision variables can be assigned an arbitrarily large value (perhaps infinity), then a solution is considered *unbounded* (Sierksma, 1996). And, if there are two or more constraints which act to preclude the assignment of any value to a decision variable, then the feasible region of the solution space is considered *infeasible and empty* (Sierksma, 1996). For example, if one constraint suggests that stand 1 needs to be assigned a final harvest after time period 10, and another constraint suggests that stand 1 needs to be assigned a final harvest before time period 5, and scheduling both harvests is not possible based on the growth of trees in stand 1, the solution to the problem would be infeasible.

3.4 Slack and Dual Prices

One formal definition of *slack* is that it represents something that is not tight, and therefore it is loose within its surroundings. When used in conjunction with linear programming, slack refers to the tightness of a constraint. Each constraint has a right-hand side value. If the total amount of the right-hand side value of a single constraint has been consumed through the assignment of activities to decision variables, there would be no slack associated with the constraint. Conversely, if the amount of the right-hand side value of a constraint has not been totally consumed after a problem has been solved (after optimal assignments have been made to the decision variables), the constraint would be characterized as having a certain amount of slack.

$$\text{Slack}_{\text{constraint}} = (\text{right} - \text{hand side value}_{\text{constraint}} - \text{contributions related to the constraint through assignments to decision variables})$$

Feasibility is maintained when a solution to a problem consumes all or less of the right-hand side value of a constraint. Once a problem has been solved, if a constraint has no (0) slack, it is considered *binding* on the solution. In a way, binding constraints control the quality of the solution to the formulated problem. Knowledge of the binding constraints can be important in understanding where the opportunities to improve the objective function value might lie (e.g., by providing more of a binding constraint's right-hand side value, if possible). For binding constraints, linear programming algorithms typically also provide a *dual price*, which reflects the amount that the objective function value would increase if one more unit of the right-hand side value of a binding constraint were available. This information could also be of value in assessing alternative courses of action to the optimal solution to a linear programming problem. Knowing how much one extra unit of a right-hand side value could contribute to improving the objective function value to a problem might prompt action (i.e., acquire one more unit through purchase, trade, etc.) on behalf of a landowner or decision-maker.

As slack and dual prices relate to the problem we have been following in this chapter, one can see below that constraints 1–3 (Rows 2–4) and constraint 7 (Row 8) have some slack. With respect to Fig. 2.9, these quantities can be seen in the third column from the right, and they are positioned next to (second column from the right) the associated variables. There are no dual prices associated with these constraints; however, the other constraints (4–6 on Rows 5–7) are binding, and the dual prices indicate how much improvement in the objective function value might be realized if one more unit of the right-hand side value of these constraints were made available to the problem. For example, if B + C ≤ 80 (row 7 of the detached coefficient matrix) were changed to B + C ≤ 81, the objective function value should increase from 1948 to 1962.

Variable	Value	Reduced Cost
A	28.00000	0.000000
B	42.00000	0.000000
C	38.00000	0.000000
TOTAL	108.0000	0.000000

Row	Slack or Surplus	Dual Price
1	1948.000	1.000000
2	7.000000	0.000000
3	13.00000	0.000000
4	2.000000	0.000000
5	0.000000	9.000000
6	0.000000	3.000000
7	0.000000	14.00000
8	616.0000	0.000000
9	0.000000	0.000000

3.5 Reduced Costs

Reduced costs are another metric provided through the output of linear programming. They do not reflect the fact that something has been marked down from some standard sale price, but in a way, they reflect an amount that might be needed to make a sale. Any decision variable that was not included in the final linear programming solution will have a reduced cost. This is the amount that a decision variable's coefficient value in the objective function would have to increase before that decision might become a part of the final solution.

Example

In the problem we have been following thus far in this chapter, none of the variables have a dual price in the optimal solution to the problem. Therefore, to demonstrate the utility of the dual price, consider the problem formulation below. Here there are three choices for a stand of trees (stand 1): cut the stand in year 1, year 2, or year 3. The decision variables (S1Y1, S1Y2, and S1Y3) are designed to contain hectares. According to the first constraint, a resource constraint, there are only 100 hectares available.

```
Maximize
4025 S1Y1 + 4127 S1Y2 + 4241 S1Y3
Subject to
S1Y1 + S1Y2 + S1Y3 <= 100
S1Y1 >= 50
S1Y2 + S1Y3 <= 60
```

In the objective function, the coefficients represent dollars per hectare for harvesting the stand in each of the three time periods. There is no constraint that indicates the stand must be harvested completely in a single year. In fact, at least 50 hectares must be harvested in year 1 (the second constraint, a policy constraint), and no more than 60 hectares can be harvested in years 2 and 3 combined (the third constraint, again a policy constraint). The solution to the problem indicates (solved using LINGO 20.0.18 (LINDO Systems Inc., 2023)) that 50 hectares will be harvested in year 1, and the other 50 hectares will be harvested in year 3. The objective function value of this optimal solution is $413,300.

```
Objective value: 413300.0
Elapsed runtime seconds: 0.04
```

(continued)

```
Variable Value Reduced Cost
S1Y1 50.00000 0.000000
S1Y2 0.000000 114.0000
S1Y3 50.00000 0.000000

Row Slack or Surplus Dual Price
1 413300.0 1.000000
2 0.000000 4241.000
3 0.000000 -216.0000
4 10.00000 0.000000
```

As you may see, the decision variable that is not included in the final, optimal solution (S1Y2) has a reduced cost of 114 (dollars per hectare). If the value of S1Y2 in the objective function increases by \$114 per hectare, to \$4241, it might force S1Y2 (cutting stand 1 in year 2) into the final solution. In this case, *might* is the appropriate word since this coefficient (4241) is the same as the value for S1Y3. However, if the value of S1Y2 increases to 4241.01, it *certainly* would enter into the final solution to this problem.

Tying this discussion back to the previous discussion concerning dual prices, you can see by the results that if one more unit of the right-hand side value of constraint 1 (S1Y1 + S1Y2 + S1Y3 < = 100, as represented by Row 2) was available, the objective function would increase by \$4241. However, if one more unit of the right-hand side value of constraint 2 was *available* (S1Y1 > = 51), the objective function would decline by \$216, as suggested by the dual price. This occurs because constraint 1 (Row 2) limits the size of the stand to 100 ha, and constraint 2 forces S1Y1, an inferior choice for this stand, to be represented by an additional hectare. By enforcing this change, the optimal solution takes 1 ha away from the better choice for the stand (e.g., S1Y3). Therefore, by taking away 1 ha from S1Y3, and adding 1 ha to S1Y1, the objective function value will decline, which is what the dual price effectively suggests.

3.6 Dual Problem

With respect to linear programming, there are two problem formulations: the original formulation emulating a real-world system, as described by an objective function and constraints (the *primal problem*), and a formulation that provides a different perspective on the problem, where the coefficients and variables are reversed (the *dual problem*).

- A primal problem can be converted to a dual problem.
- A dual problem can be converted back to the primal problem.
- Both problems should arrive at the same objective function value.

Consider the following primal problem:

```
Max 1000 X1 + 500 X2
subject to
4 X1 + 1.5 X2 <= 24
240 X1 + 30 X2 <= 1200
20 X1 + 20 X2 <= 200
```

For this problem that contains only two variables, the feasible region of the solution space (Fig. 2.10) is bound by the three constraints and the two axes.

The dual problem is the transpose of the primal problem, and the objective shifts from maximization to minimization (or vice versa should the primal have been a minimization problem) (Dantzig et al., 1956).

```
Min 24 A + 1200 B + 200 C
subject to
4 A + 240 B + 20 C  >= 1000
1.5 A + 30 B + 20 C  >= 500
```

For example, in Fig. 2.11, it becomes apparent that the objective function coefficients for the two variables in the primal problem become the right-hand side constraint values of the dual problem.

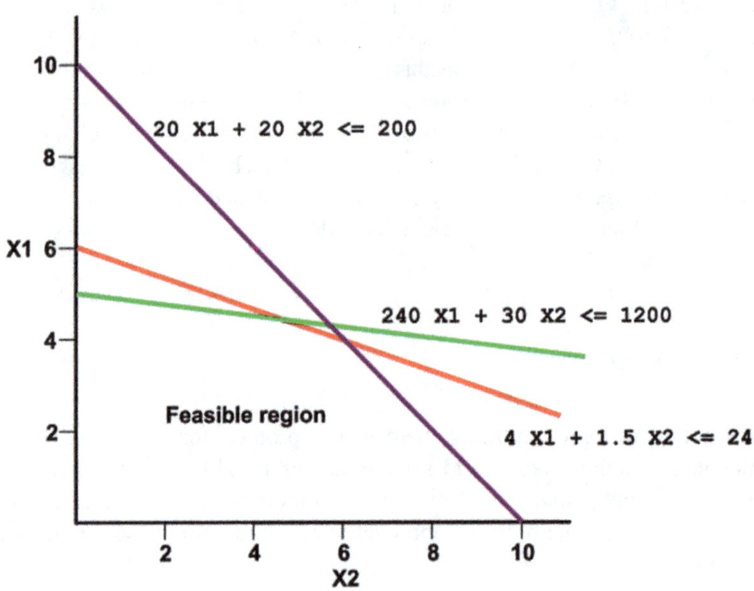

Fig. 2.10 The feasible region of a small linear programming problem. Credit: Pete Bettinger

Primal

	X1	X2		Right hand side
1)	1000	500		
2)	4	1.5	<=	24
3)	240	30	<=	1200
4)	20	20	<=	200

Dual

	A	B	C		Right hand side
1)	24	1200	200		
2)	4	240	20	>=	1000
3)	1.5	30	20	>=	500

Fig. 2.11 Transposing the objective function values of the primal problem to the constraints of the dual problem. Credit: Pete Bettinger

Primal

	X1	X2		Right hand side
1)	1000	500		
2)	4	1.5	<=	24
3)	240	30	<=	1200
4)	20	20	<=	200

Dual

	A	B	C		Right hand side
1)	24	1200	200		
2)	4	240	20	>=	1000
3)	1.5	30	20	>=	500

Fig. 2.12 Transposing the constraint right-hand side value of the primal problem to the objective function of the dual problem. Credit: Pete Bettinger

Primal

	X1	X2		Right hand side
1)	1000	500		
2)	4	1.5	<=	24
3)	240	30	<=	1200
4)	20	20	<=	200

Dual

	A	B	C		Right hand side
1)	24	1200	200		
2)	4	240	20	>=	1000
3)	1.5	30	20	>=	500

Fig. 2.13 Transposing the technical coefficients related to one variable in the primal problem to the coefficients of a single constraint in the dual problem. Credit: Pete Bettinger

Similarly, the right-hand side constraint values of the primal problem become the objective function coefficients for the dual problem (Fig. 2.12). Thus, the number of decision variables in the primal problem defines the number of constraints in the dual problem, and the number of constraints in the primal problem defines the number of decision variables in the dual problem. Notice as well that the equivalence signs related to the constraints have changed.

The technical coefficients for one of the variables, across all of the constraints in the primal problem then become the technical coefficients for a single constraint in the dual problem (Fig. 2.13). The symmetry between the primal and dual problems is further illustrated in the conceptual model displayed as Fig. 2.14.

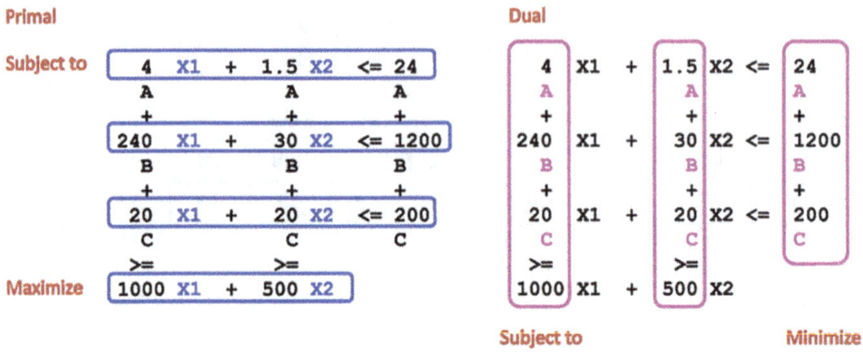

Fig. 2.14 Symmetry between the primal problem and the dual problem. Credit: Pete Bettinger

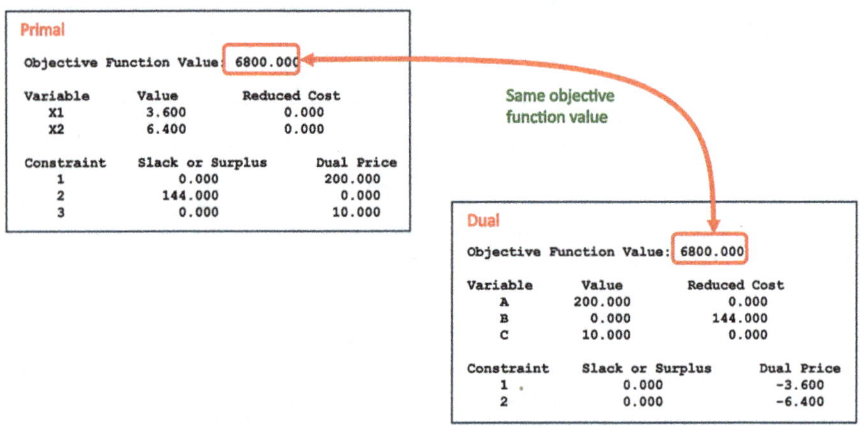

Fig. 2.15 Solutions to both the primal and dual problems, indicating equal objective function values. Credit: Pete Bettinger

Results of solving both the simple primal and dual problems presented here using LINGO 20.0.18 software (LINDO Systems Inc., 2023) indicated, as suggested, that they have the same objective function value (Fig. 2.15). Further, the values assigned to the decision variables in the primal problem are related to the dual prices of the constraints associated with the dual problem (Fig. 2.16). In a very similar manner, the values assigned to the decision variables in the dual problem are related to the dual prices of the constraints associated with the primal problem (Fig. 2.17). Further, the slack values for constraints in the primal models are related to the reduced costs of the variables in the dual model and vice versa (Fig. 2.18).

To summarize, should the primal problem represent a maximization of some objective, the dual problem will represent the minimization of some objective. If a feasible solution to one of the two problems can be located, a feasible solution to the other should also be located. Similarly, if the optimal solution for one of the two

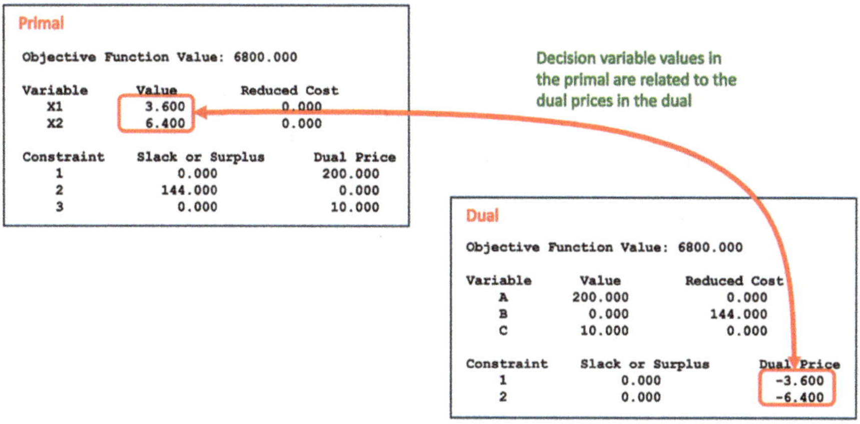

Fig. 2.16 Relationship between the values assigned to decision variables in the primal problem and the dual prices of the constraints in the dual problem. Credit: Pete Bettinger

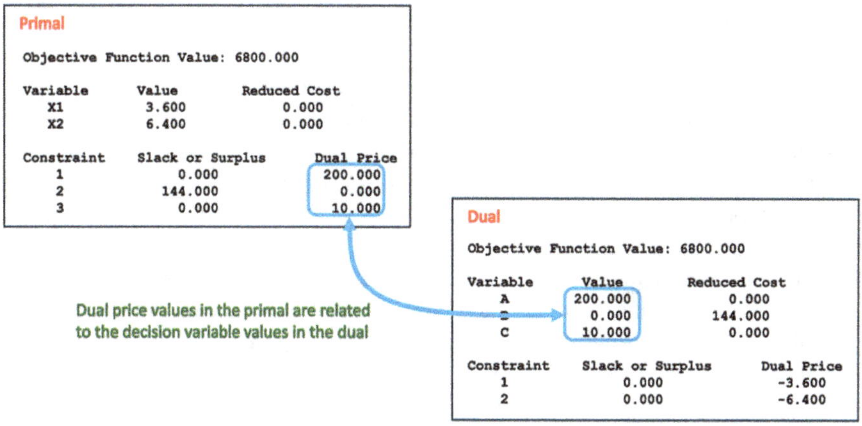

Fig. 2.17 Relationship between the values assigned to decision variables in the dual problem and the dual prices of the constraints in the primal problem. Credit: Pete Bettinger

problems can be located, the optimal solution to the other should also be located, and the objective function values should be the same. Since the objective function value of the primal and dual problems are equal, and the dual problem provides optimality criteria for the primal problem (Sierksma, 1996). Finally, as was noted earlier, it should be obvious that the dual problem of the dual problem is once again the primal problem. In sum, characteristics of the relationship between a primal problem and a dual problem include the following:

- The dual problem includes one decision variable for each constraint represented in the primal problem.

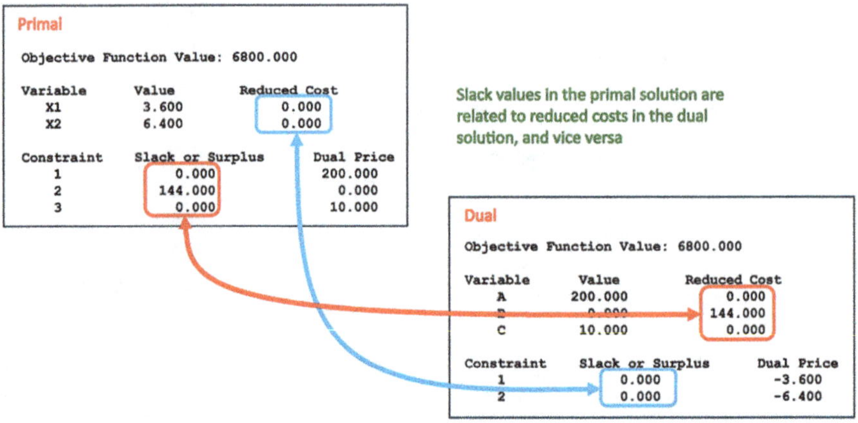

Fig. 2.18 Relationship between the slack values of the constraints and the reduced costs of the decision variables. Credit: Pete Bettinger

- The dual problem includes one constraint for each decision variable represented in the primal problem.
- The objective coefficient of the kth variable in the dual problem is equal to the right-hand side value of the kth constraint in the primal problem.
- The right-hand side value of the *k*th constraint in the dual problem is equal to the objective function value of the *k*th decision variable in the primal problem.
- The technical coefficient associated with constraint i, variable j in the dual problem is equal to the technical coefficient associated with constraint j, variable i in the primal problem.
- If there are β decision variables and ω slack variables in the primal problem, there should be ω decision variables and β slack variables in the dual problem.
- Only one of the following statements regarding the primal and dual problems is valid:

  An optimal solution can be located for both the primal and the dual problems.
  One of the two problems has an unbounded solution, and the other has an infeasible solution.
  Both of the problems have infeasible solutions.

Finally, we should note that in transforming a primal problem to a dual problem, the constraints containing inequalities may need to be adjusted prior to the transformation process, so that they all have the same characteristic. For example, in maximization problems, the constraints of the primal problem should all possess *less than or equal to* relationships with the right-hand side values. The following two example constraints from a maximization problem

```
4 X + 2 Y >= 5
3 X - 2 Y = 1
```

should therefore be adjusted to reflect constraints with *less than or equal to* relationships with the right-hand side values. The first of these is simply adjusted so that

```
-4 X - 2 Y <= -5
```

Interestingly, the constraint containing the strict equality relationship,

```
3 X - 2 Y = 1
```

can be described with two inequality constraints.

```
3 X - 2 Y <= 1
3 X - 2 Y >= 1
```

The former possesses the condition we seek, while the latter does not. Therefore, the latter is further adjusted to

```
-3 X + 2 Y <= -1
```

The computational difficulty in solving a linear programming problem is approximately proportional to the number of constraints and decision variables within the problem.

$$\text{Computational difficulty} = (\text{constraints})^2 \times (\text{decision variables})$$

In the primal problem described above, where there were three constraints and two decision variables, and assuming that there are slack variables associated with each constraint, the computational difficulty might be evaluated or scored as

$$\text{Computational difficulty} = (3)^2 \times (5) = 45$$

The associated dual problem only contained two constraints, yet it too would have five decision variables when the slack variables are added to the problem. Therefore, the computation complexity of this problem might be scored as

$$\text{Computational difficulty} = (2)^2 \times (5) = 20$$

In theory, therefore, the optimal solution to a dual problem might be computationally easier to locate when there are many more constraints than decision variables in the primal problem.

4 Contemporary Forest Harvest Scheduling Problem

A small forest management problem is described in this section to further emphasize the usefulness of linear programming to forest harvest scheduling efforts. The forest is located in southern Arkansas in the southern region of the United States (Fig. 2.19). Thirty-one management units (stands) are contained in this 872.7 ha (2156.5 ac) forest, and they are mostly composed of southern pines and deciduous tree species typical of the eastern United States. The management problem is somewhat simple for this example: the landowner desires to maximize the net discounted revenue from the harvests of trees, subject to constraints that (a) limit the harvest so that each stand is potentially only harvested once during the 15-year time period, (b) limit the total harvest volume each year to 19,000 tons (where there are 2000 pounds (907 kg) per ton), an amount which the landowner believes is sustainable over a long period of time, and (c) ensure that the total harvest volume is the same each year, to provide a stable revenue source for the landowner.

Potential wood volumes are described in tons per unit area (tons per acre). The stumpage price (the amount that the landowner receives for their wood) assumed for pine products is \$20 United States dollars (USD) per ton and the stumpage price assumed for deciduous wood products is US\$23 USD per ton. Final harvests (clearcuts) are the only management activities assumed in this harvest scheduling exercise, and the minimum final harvest age is 18 years. Reforestation is assumed to occur in the same year as final harvest activities. The reforestation (site preparation and planting) costs assumed are US\$642.46 per hectare (US\$260 per acre). Finally, the discount rate assumed is 6%.

Certainly, the management problem for the model system can be specified in more detail, but for the purposes of this section of the chapter, simplicity was desired. For example, the wood produced from each stand each year could have been more precisely described by potential products (sawtimber, pulpwood, etc.). The stumpage prices may then have been assumed to vary based on the proportion of each product present in each stand. Further, the stumpage prices are constant throughout the 15-year time horizon, which does not account for price inflation or any potential market fluctuation that may be of concern to the landowner. And, additional management practices (thinning, fertilization, and others) are ignored, as is a potential lag in time (1–2 years) between the harvest activity and the reforestation effort.

For this model system, the management problem is one where the potential discounted net revenue for each stand is maximized. Here, decision variables are noted with the symbols *SxPz*, where x = stand number and z = time period or year of harvest. Thus, the variable *S1P1* represents the scheduling of stand 1 for harvest during time period 1 (year 1). There are potentially 465 decision variables associated with this problem (31 stands × 15 time periods), although some decision variables that are associated with stands younger than the minimum harvest age each year might be removed from the problem. For example, in the abbreviated objective function below, stand 31 contributes no value to the objective until time period

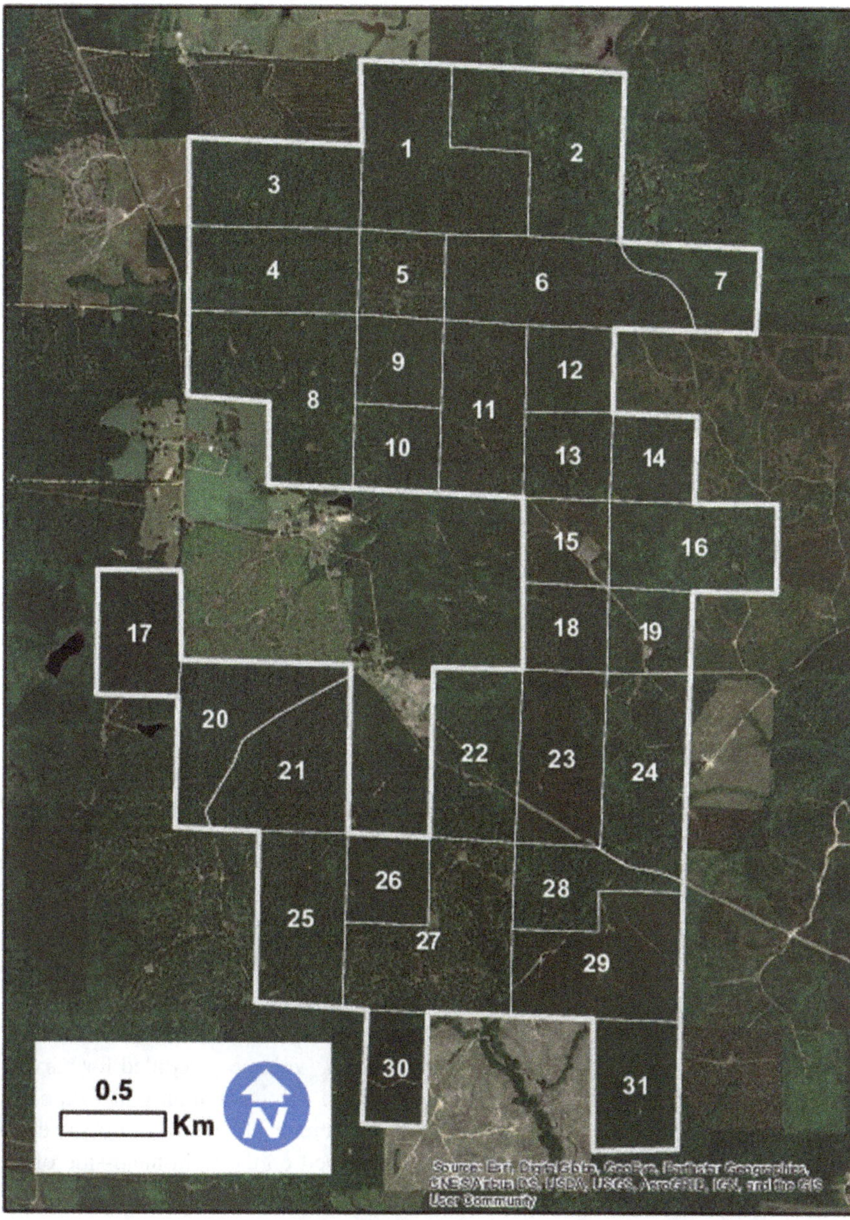

Fig. 2.19 A hypothetical forest located in southern Arkansas (United States). Credit: Pete Bettinger

9 (year 9), when at that point it reaches 18 years of age. The coefficients noted beside each variable in the objective function are the total net discounted revenue associated with a final harvest management activity.

```
Maximize
318808.87 S1P1 + 313359.15 S1P2 + 306872.63 S1P3 + 299955.48 S1P4 +
292070.00 S1P5 + 283779.09 S1P6 + 275332.00 S1P7 + 266459.46 S1P8 +
257404.97 S1P9 + 248408.31 S1P10 + 239315.76 S1P11 + 230219.04 S1P12 +
221144.39 S1P13 + 212253.69 S1P14 + 203561.19 S1P15 +
...
0.00 S31P1 + 0.00 S31P2 + 0.00 S31P3 + 0.00 S31P4 + 0.00 S31P5 + 0.00 S31P6 +
0.00 S31P7 + 0.00 S31P8 + 46793.68 S31P9 + 48989.33 S31P10 + 50529.01 S31P11 +
51555.26 S31P12 + 52131.57 S31P13 + 52369.37 S31P14 + 52260.26 S31P15
```

Resource constraints are added to the problem to prevent the linear programming process from assigning an infinite (∞) amount of harvest to each decision. In this problem, each stand can only be harvested once during the 15-year time horizon. Since the coefficients in the objective function represent total potential net discounted revenue from a final harvest, the variables can only range from 0 (no harvest activity) to 1 (complete harvest of the stand). Therefore, constraints are devised to ensure, of the 15 potential choices of activities for each stand, that at most, these choices allow only one complete harvest during the 15-year time horizon.

```
S1P1 + S1P2 + S1P3 + ... + S1P15 <= 1
S2P1 + S2P2 + S2P3 + ... + S2P15 <= 1
...
S31P1 + S31P2 + S31P3 + ... + S31P15 <= 1
```

Notable here is the fact that each stand does not *need* to be harvested. The sum of the harvest decisions can be 0, indicating no harvest is scheduled for a stand during the 15-year time horizon.

To control the wood flow, or volume of wood scheduled for harvest, some accounting rows are needed to determine the volume scheduled for harvest each year. The following accounting row multiplies the decision for each stand during time period 1 (again with a range of 0–1) by the total amount of wood (coniferous and deciduous) that is present in each stand during time period 1. The summation of these volumes is then contained in the variable *VC1* (volume scheduled for harvest during time period 1). Some stands (9, 12, etc.) have no potential harvest volume in time period 1 due to the minimum harvest age requirement. The accounting row ends with "`- VC1 = 0`" because a variable is not allowed on the right-hand-side of an equation in a linear programming problem formulation.

```
17882.436 S1P1
+ 19515.937 S2P1
+ 12888.876 S3P1
+ 12903.824 S4P1
+ 6345.253 S5P1
+ 14078.877 S6P1
+ 4649.268 S7P1
+ 16078.724 S8P1
+ 0.000 S9P1
+ 4823.465 S10P1
+ 9886.590 S11P1
+ 0.000 S12P1
+ 0.000 S13P1
+ 0.000 S14P1
+ 0.000 S15P1
+ 13584.497 S16P1
+ 9156.405 S17P1
+ 6341.340 S18P1
+ 5580.877 S19P1
+ 13152.503 S20P1
+ 16019.812 S21P1
+ 8973.460 S22P1
+ 0.000 S23P1
+ 11661.589 S24P1
+ 11905.651 S25P1
+ 0.000 S26P1
+ 19960.429 S27P1
+ 7666.659 S28P1
+ 0.000 S29P1
+ 0.000 S30P1
+ 0.000 S31P1 - VC1 = 0
```

To ensure that the scheduled harvest volumes are the same throughout the 15-year time horizon, policy constraints can be devised. Here, the potential harvest volumes in time periods 2–15 are each compared to the potential harvest volumes in time period 1, and this set of equations ensures that they all must equal the amount of wood volumes scheduled for harvest during time period 1 (hence all 15 time periods should have the same amount of wood volume scheduled for harvest).

```
VC1 - VC2 = 0
VC1 - VC3 = 0
VC1 - VC4 = 0
VC1 - VC5 = 0
VC1 - VC6 = 0
VC1 - VC7 = 0
```

```
VC1 - VC8 = 0
VC1 - VC9 = 0
VC1 - VC10 = 0
VC1 - VC11 = 0
VC1 - VC12 = 0
VC1 - VC13 = 0
VC1 - VC14 = 0
VC1 - VC15 = 0
```

In order to limit the total harvest volume each year to 19,000 tons, only one policy constraint is necessary.

```
VC1 = 19000
```

What this suggests is that if time period 1 has a scheduled harvest volume of 19,000 tons, then all other time periods would also have scheduled harvest volumes of this amount.

Using the LINGO 20.0.18 software program (LINDO Systems Inc., 2023), abbreviated results for this problem formulation are as follows:

```
Global optimal solution found.
Objective value:                    3597228.
Infeasibilities:                     0.000000
Total solver iterations:                  211
Elapsed runtime seconds:                 0.55

Variable        Value      Reduced Cost
   S1P1     0.000000        4760.636
   S1P2     0.000000        2035.337
   S1P3     0.000000        1082.687
   S1P4     0.000000        692.9622
   S1P5     0.000000        474.7788
   S1P6     0.000000        277.4593
   S1P7     0.000000        83.58930
   S1P8     0.000000        23.07350
   S1P9    0.1785568        0.000000
  S1P10     0.8214432        0.000000
  S1P11     0.000000        120.4600
  S1P12     0.000000        337.7505
  S1P13     0.000000        594.4839
  S1P14     0.000000        837.8629
  S1P15     0.000000        1169.139
   ...
    VC1     19000.00        0.000000
```

```
VC2     19000.00      0.000000
VC3     19000.00      0.000000
VC4     19000.00      0.000000
VC5     19000.00      0.000000
VC6     19000.00      0.000000
VC7     19000.00      0.000000
VC8     19000.00      0.000000
VC9     19000.00      0.000000
VC10    19000.00      0.000000
VC11    19000.00      0.000000
VC12    19000.00      0.000000
VC13    19000.00      0.000000
VC14    19000.00      0.000000
VC15    19000.00      0.000000

Row  Slack or Surplus   Dual Price
 1    3597228.          1.000000
 2    0.000000          5526.510
 3    0.000000          35584.62
 4    0.000000          4265.682
 5    0.000000          6061.125
 6    0.000000          1974.945
 7    0.000000          5445.573
 8    0.000000          7295.244
 9    0.000000          7857.211
 10   0.000000          1511.904
 11   0.000000          381.9073
 12   0.000000          698.8742
 13   0.000000          1266.822
 14   0.000000          1740.994
 15   1.000000          0.000000
 16   1.000000          0.000000
 ...
 77   0.000000          182.5376
```

These results suggest the following:

- The objective function value of this problem is $3,597,228.
- 211 iterations of the Simplex Method were used to locate the optimal solution.
- LINGO 20.0.18 (LINDO Systems Inc., 2023) solved the problem in less than 1 s.
- The total wood volume scheduled for harvest each year was 19,000 tons.
- Stand 1 (and others) was scheduled for harvest in time period 9 (17.9% of the stand) and time period 10 (82.1% of the stand). Given the divisibility assumption of linear programming, this splitting of the decision between two (or more) different time periods was possible. The wood flow constraints forced these

types of decisions. Without the wood flow constraints, the harvest decision for each stand would have been limited to a single time period (the period with the greatest value in the objective function).

- The reduced costs indicate how much each choice (by way of decision variable) would have to increase in the objective function before they would have been included in the solution to the problem. For example, the option for potentially harvesting stand 1 during time period 1 (rather than time periods 9 and 10) would have to have been $4760.64 higher in the objective function, or $323,569.51 ($318,808.87 + $4760.64). In cases where a variable was assigned a value other than 0, reduced costs are provided, otherwise, the reduced costs are 0.
- Some of the resource constraints (e.g., Rows 15 and 16 of the results) indicate some slack is present in the final solution. Here, the right-hand side values of the resource constraints were 1, and these constraints were devised to limit the final harvest to one time (1) during the 15-year time horizon. Therefore, the slack indicated in Rows 15 and 16 suggests that whichever stands were associated with those constraint rows were not harvested at all during the 15-year time horizon.
- The dual prices for the portion of results represented above may not make too much sense to this problem. They indicate how much the objective function might increase if one more unit of the right-hand side value were available for each constraint. For example, if one more unit of the constraint associated with Row 2 of the problem were available (suggesting stand 1 might be harvested twice during the time horizon), the objective function might increase by $5526.51. However, this policy (harvesting a stand twice during the time horizon) was not permissible.
- From the results, it can be seen that rows (constraints) with no slack (which are incidentally binding) are provided a dual price. Rows that contain slack are not provided a dual price.

5 Applications

Early efforts in applying linear programming to forest management problems generally assumed that the land areas were delineated as strata (collections of acres or hectares of the same type of forest), perhaps to limit the number of decision variables in the models that were described. Curtis (1962) illustrated the development of problem formulation for the management of 22,000 acres (8903 ha) of pine forests in north Florida. Two types of objective functions were described, one that maximized wood production and another that maximized the net present value of managing the forest. Constraints were devised to represent activity restrictions imposed through a management agreement with the landowner. The management costs which were recognized included those related to road management, silvicultural activities, and taxes. Wardle (1965) further illustrated the development of a problem formulation for the management of coniferous, deciduous, and mixed forests in an 8500 acre (3440 ha) area of England. The objective function of this linear programming problem was designed to maximize total discounted net revenues from management activities. The constraints involved limits on the resource available for harvesting activities, and limits on the budget allowed for reforestation activities.

In utilizing linear programming methods, the Timber Resources Allocation Method (Timber RAM) program was employed over five decades ago to determine an optimum forest harvest schedule while also controlling the volume or the area harvested (Navon, 1971; Hennes et al., 1971). One of its main uses was to determine the maximum sustainable harvest levels for a forest (Iverson & Alston, 1986). The Multiple Use-Sustained Yield Calculation technique (MUSYC) was another linear programming alternative that was developed around the same time in an attempt to improve upon the ability of Timber RAM to model certain wood flow constraints and silvicultural practices. The ideas for Model I and Model II arose from the activities to create MUSYC (Iverson & Alston, 1986). FORPLAN evolved from Timber RAM and other computerized planning systems as a linear programming method that could analyze complex forest resource interactions, accommodating all types of land and water resources in a way that assigned prescriptions (sequences of management activities) to decision variables (Bare & Field, 1986; Iverson & Alston, 1986). In the 1980s, FORPLAN was widely used to help develop US national forest plans, assisting in the exploration of potential forest plan alternatives (Bare & Field, 1986). Some planners viewed the use of FORPLAN as a process for modeling strategic planning alternatives that had tactical planning implications, while others viewed the integration of the two scales of planning more tightly within FORPLAN (Kent et al., 1991).

The use of linear programming has also been of recent interest in the assessment of alternative forest policies that consider commodity production goals along with goals for ecosystem services such as carbon stocks, water quality, and recreation values (Corrigan & Nieuwenhuis, 2016). As one example, Nahorna (2020) modeled the management of a 36,472 acre (14,760 ha) joint management area (376 landowners) in northwestern Portugal with 1343 management units (stands) containing maritime pine (*Pinus pinaster*), eucalyptus (*Eucalyptus globulus*), and other deciduous tree species. The planning problem had a 90-year time horizon with nine 10-year log time periods. Further, over a quarter million stand-level management regimes were modeled to provide data (technical coefficients) for the analyses that followed. Different objective function values were employed, such as the minimization of costs, the maximization of net present value, and the maximization of ending inventory value. In addition, constraints on wood flow, and wood production targets for different tree species and products derived from these were employed. Impacts of different management policies were then assessed with regard to management plan value, wood production levels, carbon stocks, biodiversity, and cultural values. In another recent example intended to integrate carbon dynamics into a management decision-making process, Marques et al. (2024) describe a management problem in northwest Portugal involving the same 36,472 acre (14,760 ha) joint management area. Here, a Model I resource allocation problem involving 1406 stands and 194 management prescriptions per stand (on average) was designed as a linear programming problem that accounted for wood production, carbon harvested from vegetation (trees and shrubs), fire resistance, and soil loss due to erosion. The time horizon for this problem was 100 years, which was divided into ten 10-year time periods. Three different objective functions were assessed, each was based on wood production, standing carbon stocks, and the amount of carbon removed. In some scenarios, constraints were designed to ensure a minimum amount of wood produced and a minimum fire resistance value. These examples suggest that while linear programming may be fairly mature and have characteristics (e.g., divisibility) that can concern planners, it remains a very powerful tool for assessing trade-offs, at the forest level among different management directives.

6 Summary

Linear programming can provide the optimal solution to a management problem that is appropriately designed for a model system. The Simplex Method process can be computationally cumbersome, but it is very efficient in solving problems described using linear functions. As for the computational methods illustrated in this book, linear programming represents one of the less intensive and more direct approaches for locating optimal solutions to forest harvest scheduling problems despite the assumptions (e.g., divisibility) one must accept and the associated limitations of the process. When linear programming is used to develop a forest harvest schedule, a *Principle of Optimism* is proposed:

> *A harvest scheduling effort that utilizes strata to represent land areas, or that is subject to divisibility of decision variable assignments, can be "worked out" in real life by a field forester when a management plan is implemented.*

Enhancements to the Simplex Method and other approaches to linear programming have been proposed. For example, a *Revised Simplex* method minimizes the number of computations within each iteration of the process; in essence, it involves a process that avoids transforming the entire detached coefficient matrix during each iteration of the linear programming search process (Sierksma, 1996). And, an *interior point method* has been suggested, which operates as a steepest descent process that navigates through the center of the solution space, potentially toward the optimum solution, rather than navigating around the edges of the polytope (Karmarkar, 1984; Sierksma, 1996).

Regardless of the specific linear programming method employed, information regarding potential changes to the optimal solution can be provided through the mathematical outcomes of the search process (Wardle, 1965). Slack values associated with each constraint help one understand which of the constraints are binding and thus exert more control over the allocation of resources to decisions. Reduced costs help a forest manager understand which options may be most desirable if alternatives to the optimal plan of action are necessary. Dual prices then allow a forest manager to understand how much the objective function value can improve if one more unit of the right-hand side of a constraint equation were available. These measures of sensitivity may help guide the management of a forested area when issues of uncertainty (natural disturbance, markets, and land use) arise that may not have been recognized in the formulation of the problem for the model system.

Questions

1. The assumption of divisibility in linear programming causes fractional decisions to be made when the process is applied to forest harvest scheduling problems. What are some disadvantages to this feature of linear programming when considering a contemporary forest harvest scheduling problem?
2. What does an unbounded solution suggest when attempting to solve a forest harvest scheduling problem with linear programming?
3. In your opinion, would placing an issue of concern (e.g., the development of wildlife habitat) in the objective function of a linear programming problem formulation or in the constraints of the formulation make this concern more important for the people who implement the plan?
4. After solving a forest harvest harvest scheduling problem using linear programming, is a constraint that has no associated slack important to the model system? Why?
5. If you could rename the Simplex Method, based on how it behaves in solving a linear model, what would you call it?
6. If a forest management organization is following the harvest schedule developed using a linear programming process, and they find that harvesting a certain stand of trees indicated by the solution to the problem might not be possible due to soil conditions (e.g., the ground is too wet), how can reduced costs help them select a replacement stand to harvest?
7. When solving a forest harvest scheduling problem, how does linear programming navigate through the solution space?

References

Albers, D. J., Reid, C., & Dantzig, G. B. (1986). An interview with George B. Dantzig: The father of linear programming. *The College Mathematics Journal, 17*(4), 292–314.

Avis, D. (2015). George Dantzig: Father of the simplex method. *Bulletin of the European Association for Theoretical Computer Science, 116*, 250–253.

Bare, B. B., & Field, R. C. (1986). An evaluation of FORPLAN from an operations research perspective. In T. W. Hoekstra, A. A. Dyer, & D. C. Le Master (Eds.), *FORPLAN: An evaluation of a forest planning tool*. (General Technical Report RM-140, pp. 133–144). U.-S. Department of Agriculture, Forest Service, Rocky Mountain Forest and Range Experiment Station.

Bare, B. B., & Norman, E. L. (1969). *An evaluation of integer programming in forest production scheduling problems*. (Research Bulletin No. 847). Purdue University, Agricultural Experiment Station.

Corrigan, E., & Nieuwenhuis, M. (2016). A linear programming model to biophysically assess some ecosystem service synergies and trade-offs in two Irish landscapes. *Forests, 7*(7), Article 128.

Coutu, A. J., & Ellertsen, B. W. (1960). *Farm forestry through linear programming*. (Report No. 236-60). Tennessee Valley Authority, Division of Forestry Relations.

Curtis, F. H. (1962). Linear programming the management of a forest property. *Journal of Forestry, 60*(9), 611–616.

Dantzig, G. B. (1953). *Notes on linear programming—Part III: Computational algorithm of the revised Simplex Method*. (U.S. Air Force Project RAND Research Memorandum RM-1266). The RAND Corporation.

Dantzig, G. B., Ford, L. R., Jr., & Fulkerson, D. R. (1956). *Notes on linear programming---Part XXXI: A primal—dual algorithm*. (U.S. Air Force Project RAND Research Memorandum RM-1709). The RAND Corporation.

Dantzig, G. B., Orden, A., & Wolfe, P. (1954). *Notes on linear programming: Part I the generalized simplex method for minimizing a linear row form under linear inequality restraints*. (U.S. Air Force Project RAND Research Memorandum RM-1264). The RAND Corporation.

Hennes, L. C., Irving, M. J., & Navon, D. I. (1971). *Forest control and regulation ... A comparison of traditional methods and alternatives*. (Research Note PSW-231). U.S. Department of Agriculture, Forest Service, Pacific Southwest Forest and Range Experiment Station.

Householder, A. S. (1959). *A glossary for numerical analysis*. (ORNL-2704). Oak Ridge National Laboratory.

Iverson, D. C., & Alston, R. M. (1986). *The genesis of FORPLAN: A historical and analytical review of Forest Service planning models*. (General Technical Report INT-214). U.-S. Department of Agriculture, Forest Service, Intermountain Research Station.

Kantorovich, L. V. (1960). Mathematical methods of organizing and planning production. *Management Science, 6*(4), 366–422.

Karmarkar, N. (1984). A new polynomial-time algorithm for linear programming. *Combinatorica, 4*(4), 373–395.

Kent, B. M., Dyer, A. A., & Joyce, L. A. (1991). The utility of FORPLAN for USDA Forest Service planning: The perspective of agency personnel. In M. A. Buford (Ed.), *Proceedings of the 1991 Symposium on Systems Analysis in Forest Resources*. (General Technical Report SE-74) (pp. 300–307). U.S. Department of Agriculture, Forest Service, Southeastern Forest Experiment Station.

Koopmans, T. C. (1960). A note about Kantorovich's paper, "mathematical methods of organizing and planning production". *Management Science, 6*(4), 363–365.

Leak, W. B. (1964). *Estimating maximum allowable timber yields by linear programming*. (Research Paper NE-17). U.S. Department of Agriculture, Forest Service, Northeastern Forest Experiment Station.

Leuschner, W. A., Porter, J. R., Reynolds, M. A., & Burkhart, H. E. (1975). A linear programming model for multiple-use planning. *Canadian Journal of Forest Research, 5*(3), 485–491.

LINDO Systems Inc. (2023). *LINGO 20.0.18*. LINDO Systems Inc.

Marques, S., Rodrigues, A. R., Paulo, J. A., Botequim, B., & Borges, J. G. (2024). Addressing carbon storage in forested landscape management planning—An optimization approach and application in Northwest Portugal. *Forests, 15*(3), Article 408.

Nahorna, O. (2020). *Addressing forest ecosystem management planning concerns with linear programming. An application in Portugal*. (Master of Science thesis). Universidade de Lisboa, Instituto Superior d Agronomia.

Näslund, B. (1965). Mathematical programming under risk. *The Swedish Journal of Economics, 67*(3), 240–255.

Navon, D. I. (1971). *Timber RAM ... A long-range planning method for commercial timber lands under multiple-use management*. (Research Paper PSW-70). U.S. Department of Agriculture, Forest Service, Pacific Southwest Forest and Range Experiment Station.

O'Connor, J. J., & Robertson, E. F. (2003). *George Dantzig*. University of St. Andrews. Retrieved April 16, 2024, from https://mathshistory.st-andrews.ac.uk/Biographies/Dantzig_George/

Odifreddi, P. (2004). *The mathematical century. The 30 greatest problems of the last 100 years*. Princeton University Press.

Sierksma, G. (1996). *Linear and integer programming. Theory and practice*. Marcel Dekker, Inc.

U.S. Department of Agriculture, Forest Service. (1978). *Final environmental statement and timber management plan, Tahoe National Forest*. U.S. Department of Agriculture, Forest Service, Pacific Southwest Region.

Wardle, P. A. (1965). Forest management and operational research: A linear programming study. *Management Science, 11*(10), B260–B270.

Williams, H. P. (2005). Obituary: George B Dantzig, 1914-2005. *Journal of the Operational Research Society, 56*(12), 1349–1350.

Chapter 3
Mixed Integer Programming

Learning Objectives
The objectives of this chapter on mixed-integer programming are to help the reader

- Understand the history and motivation of the exact, mathematical search process
- Understand the general manner in which the branch and bound search process operates
- Understand the use and value of exact methods for solving integer or mixed-integer problems
- Understand the main limitations associated with the use of mixed-integer programming

1 Introduction

Mixed integer programming is a general search process that is employed to locate optimal solutions to combinatorial problems where discrete decisions are needed for a certain number of the decision variables (Bare & Norman, 1969). In essence, when some decision variables need to be assigned an integer value, the assumption of divisibility in linear programming is violated. The need for integer programming models is therefore motivated by concerns that values assigned to decision variables should be discrete, whole numbers. For example, if one were maximizing the number of cows one might raise in a pasture, one might not want the answer or solution to be 5.72 cows. Given an optimal linear solution to a problem, where decision variables are assigned continuous real numeric values, rounding these to the nearest integer value might produce a solution much different than the linear solution (Odifreddi, 2004). This type of problem has been referred to as a *discrete variable programming problem* (Gomory, 1958). When only two potential values, 0 and 1, are possible, these problems are sometimes referred to as *zero-one programming problems* (Killen, 2021). The search for an optimal solution to problems of this nature is also more generally referred to as *combinatorial optimization*.

P. Bettinger, *Forest Harvest Scheduling*, Springer Texts in Business and Economics,
https://doi.org/10.1007/978-3-031-89432-9_3

2 Background

Dantzig (1958a) was perhaps one of the first to describe issues related to *mixed case* or *partial linear programming* problems where there might be *r* non-negative decision variables that would be assigned a continuous real numeric value and *n-r* non-negative integer variables (*n* representing the total number of integer and continuous decision variables in the problem). Thus, as we understand it today, mixed integer programming problems are those which contain decision variables which may be assigned integer or continuous real numeric values. Two general methods to address these problems are described here, the branch and bound method and the cutting plane method.

3 Process

Mixed integer programming problem formulations appear very similar to linear programming problem formulations with one major exception—some of the decision variables are declared to only accept integer value assignments. As an extension of linear programming, three of the principles of linear programming are valid with mixed integer programming methods for solving problems: certainty, proportionality, and additivity. The divisibility principle is relaxed, and options other than the Simplex Method are then used in addressing these problems.

3.1 Branch and Bound Method

A branch and bound method for solving mathematical problems is similar to tree search algorithms in that the search process follows a tree-like path of solutions to resolve and identify the optimal solution (Rensi & Claxton, 1972). Some of the earliest attempts to use implicit enumeration of possible solutions for integer or mixed-integer programming problems, where a tree graph structure composed of nodes (solutions) and arcs (paths to other potential solutions) is assessed, were referred to as *branch and exclude algorithms* or *additive algorithms* (Balas, 1965; Bare & Norman, 1969). Today, the branch and bound method is the most widely used method for addressing integer programming problems. A branch and bound method produces an expanding *tree* of mixed-integer linear programming solutions that *grows* toward the optimal solution of the original model (Sierksma, 1996). The branch and bound method therefore involves a deterministic, intelligent partitioning of a solution space, improving upon a feasible lower bound (in a maximization problem) until no other options are available (Michalewicz & Fogel, 2002). Inherently, this search process prevents cycling or returning to previously assessed locations within a solution space, and so it has been characterized as a type of *strong*

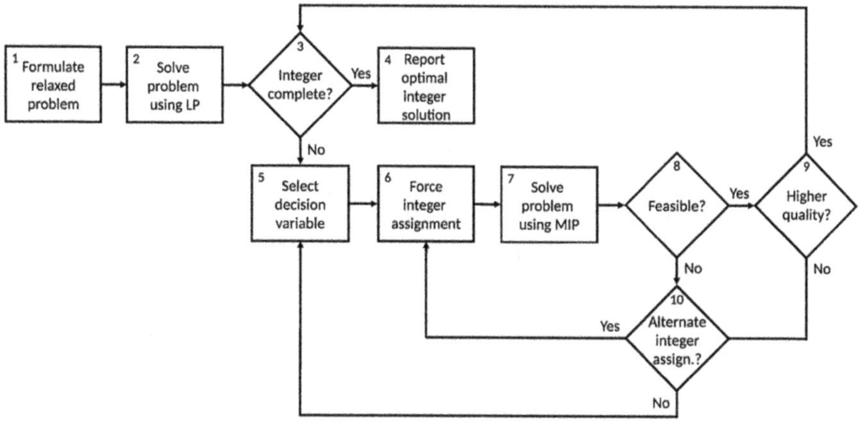

Fig. 3.1 A branch and bound process for solving an integer or mixed-integer problem. Credit: Pete Bettinger

inhibition search (Glover, 1986). The branch and bound method has also been described as a *conventional technique* for locating solutions to scheduling problems (Thalman et al., 1991). To some, it might be considered a rational method for locating a solution to a problem.

The branch and bound method is an intelligent search technique that initially solves a *relaxed linear programming model* (McDill & Braze, 2001), then examines each decision variable to locate those that were meant to have an integer value assignment yet were assigned a continuous real numeric value. One of these is selected. The possible integer values on both sides of the continuous real numeric value assignment, and the two solutions that are formed through these assignments, are identified and assessed. If either of these solutions results in an inferior objective function (as compared to the best solution saved in memory), further analysis along this branch of the search tree is terminated. This is necessary because further down the branch of the search tree, solutions will only be found with additional non-divisibility constraints that will further reduce the objective function value. If either of these solutions results in a better objective function value, this route down the branch of the search tree is further explored. The search continues until all decision variables that were meant to have an integer value actually are assigned an integer value.

A general branch and bound method might follow this course of action for a maximization problem:

- Formulate a problem with all integer restrictions relaxed (Step 1 in Fig. 3.1). An integer or mixed-integer model is considered relaxed when decision variables meant to be assigned integer values are assigned continuous real numeric values.
- Solve the relaxed problem (Step 2).
- Select a focal decision variable with a fractional assignment (Step 5).

- Force the assignment of the focal decision variable to have an integer value and re-solve the problem.

  If the temporary proposed solution is not feasible, select the alternate integer assignment for the focal decision variable (Step 10) and solve the problem once again (Step 7), or select another variable with a fractional assignment in the relaxed solution (Step 5), develop a branch, and solve the problem once again.

  If the temporary proposed solution is feasible, yet results in an objective function value that is lower in quality than the lower bound (in the case of a maximization process), select the alternate integer assignment for the focal decision variable and solve the problem once again, or discontinue search along this branch.

  If the temporary proposed solution is feasible, results in an objective function value that is higher in quality than the lower bound (in the case of a maximization process), yet does not include integer assignments in every case where an integer assignment to decision variables was required (Step 3), select one of these (Step 5) and develop a branch, and solve the problem once again.

  If the temporary proposed solution is feasible, includes integer assignments in every case where an integer assignment to a decision variable was required (Step 3), and there are no more branches to examine, terminate the search process (Step 4).

In a maximization process, the objective function value of the solution to the associated relaxed linear programming problem represents an upper bound on potential feasible solution quality. Each feasible integer solution where an integer assignment to some of the decision variables that required this designation has been made becomes the lower bound when the objective function value is higher in quality than previously observed solutions of this type. The process along a branch ends when a sub model produces an infeasible solution or it produces a solution of lower quality than the quality of the current best solution (Sierksma, 1996). Here, it is proven that continued branching will only lead to feasible solutions with inferior objective function values (Rensi & Claxton, 1972). The search process terminates when there are no more unexpanded or non-terminated nodes in the tree (McDill & Braze, 2001). When the search process locates a feasible solution with respect to all integer assignments and this solution has an objective function value higher in quality than any other node in the expanding portion of the tree, the solution represents a local optimum. If there are no other nodes to assess, this solution represents the optimal solution to the problem. If one employs an *optimality tolerance*, the search process may terminate when the difference between the relaxed upper bound and the feasible lower bound is within a pre-defined range often expressed in percentage terms (McDill & Braze, 2001). Otherwise, the search process will continue to partition (branch) the feasible region of the solution space into smaller subsets to locate potentially higher quality lower bounds.

When the search process terminates, the integer (or mixed integer) solution with the best objective function is reported. As one might gather from the description of the branch and bound method, there are two key decisions faced at each iteration of the search: which node to expand in the tree, and which variable to assign integer values to at each node (McDill & Braze, 2001). With respect to maximization problems, within the decision tree that represents the search process, each node represents a partial solution to the full problem, and at each level, the node with the best lower bound is assessed first in an attempt to locate the optimal feasible solution as quickly as possible. For a maximization problem, at each step in the process,

- There may be a provable lower bound that could be feasible.
- There is a provable, relaxed upper bound.
- We know that the optimal solution (based on the objective function value) will be found somewhere between these two solutions (and could perhaps be the solution represented by the provable lower bound).

Similarly for a minimization problem,

- There may be a provable upper bound that could be feasible.
- There is a provable, relaxed lower bound.
- We know that the optimal solution will be found somewhere between these two solutions (and could perhaps be the solution represented by the provable upper bound).

Unfortunately, the number of potential subproblems is an exponential function of the number of integer decision variables. Thus, while the branch and bound method has been characterized as practically useful, it is theoretically computationally inefficient in solving integer or mixed integer programming problems when the problem formulation consists of a large number of activities that need integer assignments (Claxton & Rensi, 1972; Sierksma, 1996). On a positive note, there are a number of options for adjusting how a branch and bound method may be employed; the combination of options selected may affect the amount of time required for the process to operate, and the best combination of options may be problem specific (McDill & Braze, 2001).

Example

As an example of the application of the branch and bound method to a *zero-one integer programming* problem, the simple formulation below contains four decision variables (*X1*, *X2*, *X3*, and *X4*) that will all require an integer assignment. Given the last four constraints in the problem, each decision variable can be assigned a value of 0 or 1.

(continued)

```
maximize 75 X1 + 6 X2 + 3 X3 + 33 X4
subject to
774 X1 + 76 X2 + 22 X3 + 42 X4 <= 875
67 X1 + 27 X2 + 794 X3 + 53 X4 <= 875
X1 <=1
X2 <=1
X3 <=1
X4 <=1
end
```

Problem 1 (Fig. 3.2) represents the outcomes of the relaxed linear programming formulation of the problem above. The objective function value is 113.837 units, and two of the decision variables (*X2* and *X3*) are assigned non-integer values. This solution represents an upper bound to the optimal solution. Of course, this solution is not the optimal solution to the problem, since it is not feasible with respect to the desired integer assignments.

Problems 2 and 7 represent branches from Problem 1, where a decision variable (*X3*) with the largest fractional value (0.934) in Problem 1 is forced to have an integer assignment below (*X3* = 0 in Problem 2) and above (*X3* = 1 in Problem 7) the value previously assigned in the outcome of Problem 1. The solution to Problem 2 has an objective function (112.658 units) value slightly below that of Problem 1, and one of the remaining decision variables (*X2*) is assigned a non-integer value. Therefore, the outcome from Problem 2 is not the optimal solution to the problem.

Two branches are then formed from Problem 2 to create Problems 3 and 4. In Problem 3, decision variable *X2* is forced to have an integer assignment below its assigned value in Problem 2 (*X2* = 0 in Problem 3) and above its assigned value in Problem 2 (*X2* = 1 in Problem 4). Notice as well that the assignment for decision variable *X3* is held constant (value = 0) in these branches of the tree. The outcome of Problem 3 indicates that all four decision variables are assigned integer values. Since this outcome represents a feasible solution to the problem, the objective function value here (108) represents a feasible lower bound. To reiterate the status of the search at this point, the upper bound is 113.837 (from Problem 1) and the lower bound is 108 (from Problem 3). No more branching is necessary from this node in the tree since there are no other decision variables with non-integer assignments.

In Problem 4, the objective function value (112.353 units) is slightly lower than the upper bound, and one decision variable (*X1*) is assigned a non-integer value (0.978). Branches from this node in the tree force decision variable *X1* to have a value of 0 in Problem 5, and a value of 1 in Problem 6. The outcome from Problem 5 is feasible, as all decision variables have an integer assignment. However, the objective function value from Problem 5 is 39 units, which

(continued)

is much lower than the feasible lower bound (108) from Problem 3. The outcome from Problem 6 provides an objective function value (100.643) which is also lower than the lower bound. Even though one decision variable (*X4*) is assigned a non-integer value in Problem 6, suggesting further branching from this node, the search process along this branch of the tree terminates since no solution below this node can contain an objective function value above 100.643 units (which is lower in quality than the known, feasible lower bound).

Returning to Problem 7, where decision variable *X3* was forced to have a value of 1 under a branch from Problem 1, the outcome here produces an objective function value of 86.717 units. Since this objective function value is lower in quality than the known lower bound (108 units), and since no other solutions produced further down the branch can produce an objective function value greater than 86.717 units, the search process terminates here.

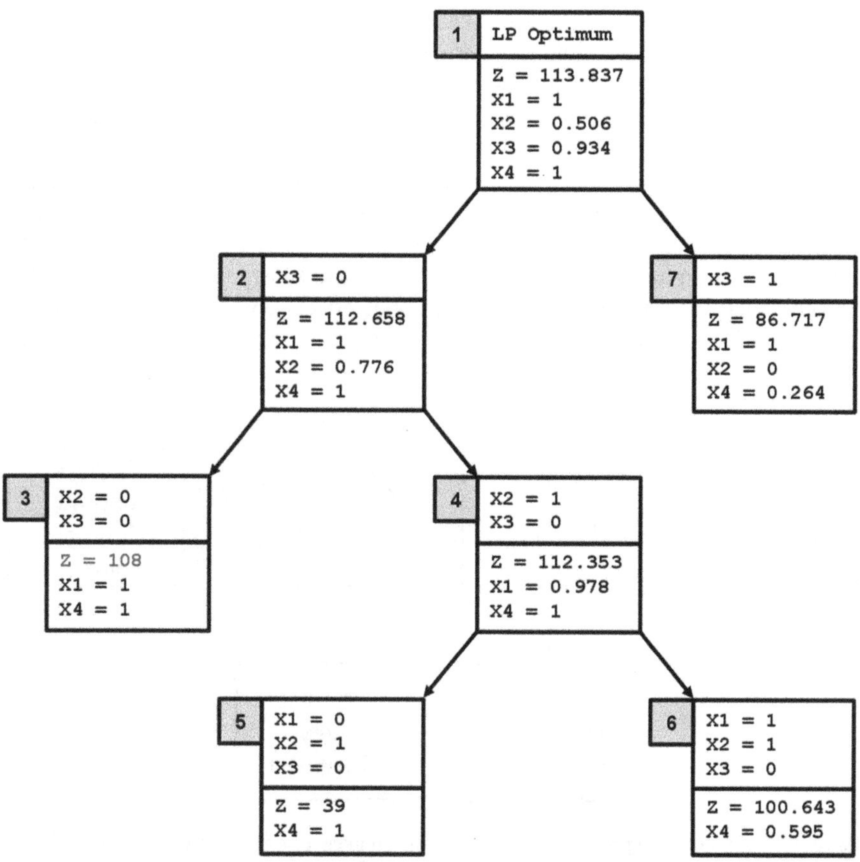

Fig. 3.2 The branching process from a small, hypothetical zero-one integer programming problem. Credit: Pete Bettinger

3.2 Cutting Plane Method

The cutting plane method is an interesting idea that involves the successive introduction of additional constraints to a linear programming model to eliminate non-integer solutions from the edges of the feasible region of the solution space (Hillier & Lieberman, 1980). In perhaps the first work describing the reduction of a solution space through the addition of hyperplanes to a linear model, and in lieu of a better name for it, Dantzig et al. (1954) simply called this process *the method*, which they admit may have been somewhat pretentious. Cutting plane methods designed for integer programming problems first solve the linear programming problem, then add a series of additional valid inequality constraints to the problem formulation to reduce (cut) the solution space while retaining all feasible integer solutions, and so that the extreme points of the solution space contain only integral solutions (Dantzig, 1958b; Wolsey, 1998). The inequality constraints that are added satisfy all integer assignment requirements yet cut off the vertex of the *n*-dimensional polytope that described the optimal relaxed linear programming solution. These have been called *cut constraints* (Sierksma, 1996). A slack variable is then associated with the cut constraint, and optimality is assessed. The process repeats until all decision variables meant to have integer variables are assigned integer variables and the dual model indicates that the optimal solution to the problem has been located.

Gomory (1958) suggested a basic conceptual cutting plane model to address mixed integer problems that is based on formulating and solving a linear problem using the Simplex Method (Steps 1 and 2 in Fig. 3.3). The resulting linear programming solution is then examined (Step 3) to determine whether integer values were assigned to decision variables where integer values were necessary. If this is not the case, a new constraint is formulated (Step 5) that will satisfy the still unknown (but desired) integer or mixed-integer solution, but effectively create a new, smaller feasible region of the solution space by cutting or trimming off regions of the

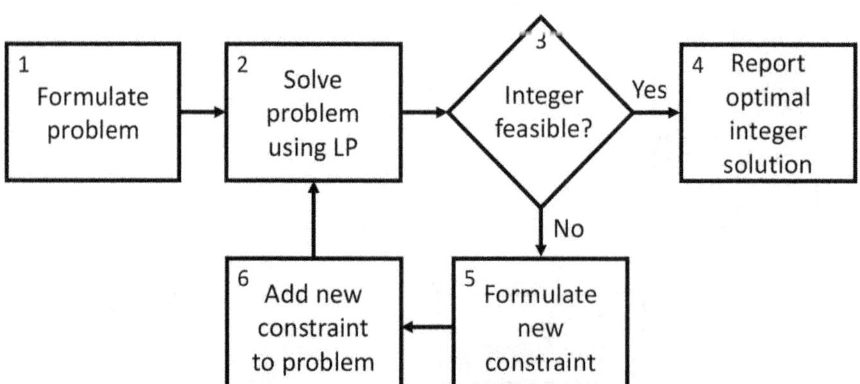

Fig. 3.3 A general cutting plane process for solving an integer or mixed-integer problem. Credit: Pete Bettinger

solution space that span non-integer assignments near one or more of the desired integer assignments to decision variables. The new constraint removes the current optimal linear programming solution from the feasible region of the solution space (Hillier & Lieberman, 1980) and passes through or includes a point representing an integer assignment to a decision variable (Dantzig et al., 1954). The new constraint is then added to the problem (Step 6) and the new problem is subject to the Simplex Method again. Typically, only one new constraint (cut) is added during each iteration of the model, but this assumption is not necessarily firm (Balas et al., 1996). The search process is repeated until an optimal solution has been obtained that contains integer assignments to decision variables that required integer assignments (Step 4).

4 Contemporary Forest Harvest Scheduling Problem

Expanding upon the small forest management problem from Sect. 4 of Chap. 2, here each of the decision variables that relate to the period of assignment of a final harvest to a stand are required to contain a binary integer value (0 = do not cut a stand during the time period suggested, 1 = harvest a stand during the time period suggested). As there are other variables in the problem that relate to the accounting rows, those that contain volumes accumulated for each time period, the resulting mathematical problem formulation is of a mixed-integer type. In addition to the assumptions that were described previously regarding prices and costs, this revised problem also includes pairwise adjacency constraints, where when one stand is scheduled a final harvest during a certain time period, another physically adjacent stand cannot be scheduled for a final harvest during the same time period.

```
S1P1 + S2P1 <= 1
```

Here, stand 1 and stand 2 are physically adjacent, and only one of them can be scheduled a final harvest during time period 1. Of course, neither of these could also be assigned a final harvest during time period 1. These pairwise constraints needed to be developed for all adjacency relationships and for all time periods. Certainly, longer, multiperiod green-up requirements could be evaluated, but for the sake of simplicity, a one-period green-up requirement is acknowledged in this example.

Adding further intrigue to the problem, the average volume scheduled for harvest across the time horizon is determined, and the periodic scheduled volumes are allowed to deviate 25% above or below the average. The strict adherence to a desired harvest volume (19,000 tons per year) that was noted in Sect. 4 of Chap. 2 is relaxed here. The average in this mixed integer problem formulation is determined using this form of equation:

```
0.06667 VC2 + 0.06667 VC2 + ... + 0.06667 V15 - AVGVC = 0
```

The coefficients (0.06667) represent the fraction one-fifteenth (1/15), and so the accumulated volumes (VC1, VC2, etc.) are effectively multiplied by 1/15 to arrive at the average. Then, the accumulated volumes (VC1, VC2, etc.) are restricted from deviating more than 25% from the average with equations such as these:

```
VC1 - 0.75 AVGVC >= 0
VC1 - 1.25 AVGVC <= 0
...
VC15 - 0.75 AVGVC >= 0
VC15 - 1.25 AVGVC <= 0
```

Using the LINGO 20.0.18 software program (LINDO Systems Inc. 2023), abbreviated results for this problem formulation are as follows:

```
Objective value:                4480085.
Objective bound:                4562402.
Infeasibilities:                 0.000000
Extended solver steps:            184498974
Total solver iterations:           2147483647

Variable        Value
   S1P1      0.000000
   S1P2      0.000000
   S1P3      0.000000
   S1P4      0.000000
   S1P5      0.000000
   S1P6      0.000000
   S1P7      0.000000
   S1P8      1.000000
   S1P9      0.000000
  S1P10      0.000000
  S1P11      0.000000
  S1P12      0.000000
  S1P13      0.000000
  S1P14      0.000000
  S1P15      0.000000
   ...
    VC1     22420.06
    VC2     28479.51
    VC3     28264.67
    VC4     28530.71
    VC5     28399.51
```

```
  VC6     28969.17
  VC7     27744.97
  VC8     22137.88
  VC9     19492.52
  VC10    19100.27
  VC11    19028.07
  VC12    20920.66
  VC13    18091.48
  VC14    18832.38
  VC15    18135.21
  AVGVC   23237.63
```

The branch and bound process was interrupted after quite some time; thus the final result may not be the optimal solution to the problem. However, these results suggest the following:

- The objective function value of this problem is $4,480,085.
- A total of 2,147,483,647 iterations of the solver were employed to up to the point of termination of the process.
- The average wood volume scheduled for harvest each year was about 23,238 tons.
- The actual wood volume scheduled for harvest each year was between 75% and 125% of the average scheduled for harvest.
- Stand 1 was scheduled for harvest in time period 8. The integer constraint forced these types of decisions.

5 Applications

In one application of mixed integer programming in forestry, Abate et al. (2024) developed a mixed integer problem formulation to assign management prescriptions to an 18,500 acre (7487 ha) forested area in Portugal that was subdivided into 686 management units (stands) containing mainly eucalypt and pine tree species. The model system was formulated as a Model I mixed-integer forest management problem. Several objectives were considered, including those that maximized wood volume scheduled for harvest, cork produced, carbon sequestered, biodiversity accommodated, and fire resistance produced, and those that minimized erosion generated from management activities. Constraints included limits on the variation in wood flow and limits on final harvest sizes (modeled as an area restriction set of constraints). The initial problem formulation included about 177,000 final harvest adjacency constraints, which prompted the subdivision of the full problem into four sub-problems that were solved separately.

An interesting application of the mixed-integer programming method was provided by Değermenci et al. (2022) to help develop a forest harvest schedule that maximized the net present value of income from timber, water, and carbon sequestration outcomes. The study area in Turkey contained 292 stands that covered 8614 acres (3486 ha). The forest was composed of beech (*Fagus orientalis*), fir (*Abies bornmulleriana*), pine (*Pinus sylvestris*, *Pinus nigra*), and hornbeam (*Carpinus betulus*) tree species, among others. Here, the planning horizon was 100 years, which was divided into five 20-year time periods. Constraints included minimum harvest ages, limits on the variation in wood flow, limits on the growth balance between successive time periods, and limits on the area to subsequently require forest regeneration activities. The inspiration for developing the planning model was to acknowledge multiple ecosystem services of the forest and incorporate them into the objective function and constraints of the problem formulation.

In an application of mixed integer programming to a very short-term tactical planning problem, Ferrari et al. (2020) describe how a Model I problem was designed to minimize the costs related to the management of a 9179 acre (3715 ha) pine (*Pinus* spp.) forest in southern Brazil. The time horizon for this problem was 5 years, and it was divided into 1-year time periods. The constraints on the harvest scheduling problem included those related to wood flow, which limited fluctuations in production from 1 year to the next, and those related to control over the dispersion of harvest locations within each year. The design of the problem was inspired by a desire to reduce the risks associated with wood production and the economics of forest management.

6 Summary

The development of the cutting plane and the branch and bound algorithms has been noted as prominent landmarks in the history of development of integer programming methods (Glover, 1986). While the most effective general-purpose mixed integer programming processes have centered on the branch and bound algorithm, the difference between the two is not necessarily distinct, as constraining integer variables to satisfy upper and lower bounds within a branch and bound algorithm can be envisioned as empowering provisional cutting practices in the cutting plane method (Glover, 1986). If a mixed integer programming search process is allowed to conclude (without interruption) an optimal solution to a combinatorial problem may be assured, assuming there is at least one feasible solution to the problem.

For contemporary forest management problems, where the adjacency (or size) of harvest activities and a green-up delay for scheduling others nearby might be important, integer decision variables have often been used to represent the management decisions. For relatively large problems, mixed integer programming search processes can be computationally slow compared to others presented in this book. The efficiency of methods that are developed to solve mixed-integer programming problems has been suggested to be a function of the size (number of decision variables and constraints) of the problem itself (Bare & Norman, 1969). Further, the effort required to develop constraints related to a maximum final harvest size is a function of the number of stands that can contribute to an opening without violating the constraint (Crowe et al., 2003). In solving these problems, the worst case scenario may involve a nearly complete enumeration of the feasible solutions within the solution space. However, the goal of mixed integer methods is not to visit and assess all of the nodes in a decision tree; some intelligent logic is employed to reduce the extent of the search. With respect to forest harvest scheduling, a *Principle of Effort* is proposed:

> *A harvest scheduling effort that is relatively easy to implement by a planner* (e.g., *using linear programming or binary search, recognizing land areas as strata, and avoiding spatial constraints) for some model system (a forest) may be relatively difficult to implement in real life by a field forester.*
>
> *Conversely, a harvest scheduling effort that is relatively difficult to implement by a planner* (e.g., *using something other than linear programming or binary search, recognizing areas as individual stands, and incorporating spatial constraints) for some model system (a forest) may be relatively easy to implement in real life by a field forester.*

Take Home Message

Mixed integer programming is a method for locating exactly the optimal solution to a problem where some of the decision variables are assigned integer values. Often, either a branch and bound or a cutting plane technique (or some combination of the two) is employed to identify the optimal solution within the solution space. With a branch and bound search process, a tree-like conceptual model is employed to determine how far down each limb the search must go before it decides that no more feasible or efficient solutions can be found. With the cutting plane search process, the solution space is reduced in size as new constraints (cutting planes) are added to the model that intersect feasible integer solutions. These capabilities allow the mixed integer programming to locate feasible and efficient solutions to complex problems, and if they are not terminated early, optimal solutions to complex problems.

Questions

1. It has been argued that mixed integer programming represents contemporary forest harvest scheduling problems better than linear programming so that, for example, harvest adjacency constraints can be modeled appropriately. However, is has also been argued that a linear programming solution is *good enough*, and that field foresters can work out any constraint-related problems of the harvest schedule during implementation of the plan of action. Where do you stand on this issue?
2. Mixed integer programming has been characterized as a *smart*, *intelligent*, *rational* alternative to a complete enumeration of all possible solutions to a complex problem. What other adjectives might you associate with this search process? Why?
3. If an optimality tolerance is assumed when solving a mixed integer programming forest harvest scheduling problem using the branch and bound methods, how large would you allow the tolerance to be (in percentage terms of the objective function value of the relaxed linear programming solution) before you feel it may

fail to locate some potential higher quality solutions to the problem because the search process terminated too early?

4. The *Principle of Effort* is related to real-life implementation of a forest harvest schedule. What caveats come to mind that might be used to relax what seems to be a dichotomous point of view?

References

Abate, D., Marques, S., Bushenkov, V., Riffo, J., Weintraub, A., Monstantino, M., Lagoa, C., & Borges, J. G. (2024). Assessment of tradeoffs between ecosystem services in large spatially constrained forest management planning problems. *Frontiers in Forests and Global Change, 7*, Article 1368608.

Balas, E. (1965). An additive algorithm for solving linear programs with zero-one variables. *Operations Research, 13*(4), 517–546.

Balas, E., Ceria, S., Cornuéjols, G., & Natraj, N. (1996). Gomory cuts revisited. *Operations Research Letters, 19*(1), 1–9.

Bare, B. B., & Norman, E. L. (1969). *An evaluation of integer programming in forest production scheduling problems.* (Research Bulletin No. 847). Purdue University, Agricultural Experiment Station.

Claxton, H. D., & Rensi, G. (1972). *An analytical procedure to assist decision-making in a government research organization.* (Research Paper PSW-80). U.S. Department of Agriculture, Pacific Southwest Forest and Range Experiment Station.

Crowe, K., Nelson, J., & Boyland, M. (2003). Solving the area-restricted harvest-scheduling model using the branch and bound algorithm. *Canadian Journal of Forest Research, 33*(9), 1804–1814.

Dantzig, G. B. (1958a). *On integer and partial integer linear programming problems.* (P-1410). The RAND Corporation.

Dantzig, G. B. (1958b). *Notes on linear programming: Part XLVII Solving linear programs in integers.* (U.S. Air Force Project RAND Research Memorandum RM-2209). The RAND Corporation.

Dantzig, G. B., Fulkerson, R., & Johnson, S. (1954). *Solution of a large scale traveling salesman problem.* (P-510). The RAND Corporation.

Değermenci, A. S., Zengin, H., & Ünal, M. E. (2022). Optimization of water yield, carbon storage and wood production in forest planning by mixed integer programming. *Bosque, 43*(1), 7–21.

Ferrari, L. S., Arce, J. E., Pelissari, A. L., da Silva, J. P., Figueiredo Filho, A., & de Oliveira, E. B. (2020). Tactical planning of forest harvesting under different scheduling restrictions. *Scientia Forestalis, 48*(127), e3334.

Glover, F. (1986). Future paths for integer programming and links to artificial intelligence. *Computers & Operations Research, 13*(5), 533–549.

Gomory, R. E. (1958). Outline of an algorithm for integer solutions to linear programs. *Bulletin of the American Mathematical Society, 64*(5), 275–278.

Hillier, F. S., & Lieberman, G. J. (1980). *Introduction to operations research* (3rd ed.). Holden-Day, Inc.

Killen, J. (2021). *Mathematical programming methods for geographers and planners*. Routledge.

McDill, M. E., & Braze, J. (2001). Using the branch and bound algorithm to solve forest planning problems with adjacency constraints. *Forest Science, 47*(3), 403–418.

Michalewicz, Z., & Fogel, D. B. (2002). *How to solve it: Modern heuristics*. Springer-Verlag.

Odifreddi, P. (2004). *The mathematical century. The 30 greatest problems of the last 100 years.* Princeton University Press.

Rensi, G., & Claxton, H. D. (1972). *A data collection and processing procedure for evaluating a research program.* (Research Paper PSW-81). U.S. Department of Agriculture, Pacific Southwest Forest and Range Experiment Station.

Sierksma, G. (1996). *Linear and integer programming. Theory and practice.* Marcel Dekker, Inc.

Thalman, N., Sparn, T., Jaffres, L., Gablehouse, D., Judd, D., & Russell, C. (1991). AI techniques for a space application scheduling problem. In J. L. Rash (Ed.), *1991 Goddard Conference on Space Applications of Artificial Intelligence.* (NASA Conference Publication 3110) (pp. 83–93). National Aeronautics and Space Administration, Office of Management, Scientific and Technical Information Division.

Wolsey, L. A. (1998). *Integer programming.* John Wiley & Sons, Inc.

Chapter 4
Goal Programming

Learning Objectives
The objectives of this chapter on goal programming are to help the reader

- Understand the history and motivation of the goal programming process
- Understand the general manner in which goal programming operates
- Understand the use of weights to equalize or favor different goals
- Understand the main limitations associated with the use of goal programming

1 Introduction

Two important aspects of multiple-use forest management are the formulation of goals and the flexibility to accommodate them through management activities (Bare & Anholt, 1976). With goal programming, a person may desire to develop a forest harvest schedule where there are conflicting objectives that have desired levels of outcomes, and each of these are expected to be simultaneously produced as closely as possible to their desired level of achievement (Killen, 2021). Goal programming requires explicit representation of quantifiable goals in the objective function and a weighting or preference structure for each to potentially bias the attainment of goals that are deemed more (or less) important to the forest manager (Field, 1973). In this regard, goal programming is considered a multiple objective procedure that is not limited to a single decision criterion (Hotvedt, 1983). Initially, goal programming was designed as a special extension of both linear and mixed-integer programming, as a useful tool for weighing multiple, potentially conflicting objectives when developing a forest harvest schedule (Lee, 1973). However, the concept of goal programming is not limited to exact mathematical approaches for solving problems, as a goal programming problem can be addressed with heuristic search processes as well.

P. Bettinger, *Forest Harvest Scheduling*, Springer Texts in Business and Economics,
https://doi.org/10.1007/978-3-031-89432-9_4

2 Background

Goal programming arose from the *goal attainment* process described by Charnes and Cooper (1961) as a method for addressing what seemed to be at that time unsolvable linear programming problems (Bare & Anholt, 1976). Goal programming therefore became a form of linear programming that allowed one to include diverse types of concerns into an objective function. Goal programming, like linear programming and mixed integer programming, attempts to locate the very best, feasible solution to a model program, yet it is structured to simultaneously come as close as possible in meeting a set of goals or goals ordered by preference or priority (Bell, 1977). Goal programming could be described as a *normative process*, which emphasizes policies or goals that should be met (Charnes et al., 1975), even though an objective function and constraints could be designed to force the development of a harvest schedule that favors certain forest policies or goal outcomes. Further, if the desired goal outcomes that may be produced as a result of a harvest schedule are transformed from their recognition in stringent constraints to their recognition within an objective function, there is also a greater chance of obtaining a feasible solution as deviations far from target goal achievement levels still result in a feasible harvest schedule.

Imagine, for example, an objective function that recognizes financial, commodity production, wildlife suitability, and spatial concerns. Each of these may be represented by a different type of value (dollars, tons, habitat units, distances) and each may have a different valid range of outcomes (e.g., 0–10,000,000 for dollars, 0–1,000,000 for tons of wood produced, 0–1 for habitat suitability, or 0–10,000 for spatial concerns). With goal programming, one would not necessarily maximize or minimize any of these directly, whatever may be the case. A goal programming formulation would, however, be designed to minimize the difference between the achievement of these (through the scheduling of activities) and some desired target for each (the goals). Assuming that every time period of a time horizon represents a point of measurement between actual (scheduled) and desired levels of the different concerns, in using this process, one would (a) need to know the desired target for every time period of the time horizon, and (b) need to know the impact of the achievement of these concerns with every change to a harvest schedule as a search process progresses. One interesting issue that needs to be addressed involves deciding whether to place each of these concerns on equal footing, either through weighting some of the outcome values in the objective function (those that represent deviations from targets) or through normalizing the outcomes so that they seem similar. For example, one could adjust the deviations of habitat suitability (normally measured on a 0–1 scale) from a target level of achievement so that they are similar in magnitude as deviations in wood flow (hundreds or thousands of tons) from a different target level of achievement.

If the weighting and normalization issues can be overcome, the ability of goal programming to accommodate different concerns in the objective function of a forest planning problem can be attractive to forest managers. In this process the age-old argument that a harvest schedule is biased when it focuses on one concern in the objective function, while relegating the other concerns to constraints, can be muted. However, some goals can be complimentary while others are competitive (Schuler et al., 1977). For the planner, the use of goal programming introduces the problem of not only defining the target achievement levels for goals, but also the problem of explaining the targets to stakeholders. Nonetheless, because of its unique structure (minimizing deviations from targets), goal programming has found an audience in some forest management organizations around the world. Some United States national forests, such as the Caribou National Forest in Idaho, have employed goal programming as a way to assess forest plan alternatives and to compare outcomes against those obtained using single-objective linear programming problem formulations (US Department of Agriculture, Forest Service, 1984a). In an assessment of proposed forest plans for the Kisatchie National Forest in Louisiana, timber, rangeland, and wildlife habitat goals were included in the objective function of goal programming problem formulations (US Department of Agriculture, Forest Service, 1984b).

3 Process

Goal programming is a process whereby one minimizes deviations between target goal achievement levels and outcomes from the scheduling of management activities that may be produced over a time horizon, subject to resource and policy constraints (Schuler et al., 1977). Goal programming is often structured as an exact method for solving problems, much like linear programming and mixed integer programming. In fact, if designed in this manner, a goal programming problem formulation may very closely resemble a linear or mixed integer programming problem formulation. However, the concept of goal programming is not limited to the use of exact methods to address harvest scheduling problems. In fact, heuristic search techniques such as simulated annealing (Chap. 9) and others can be employed in an attempt to minimize deviations between target goal achievement levels and periodic outcomes obtained from the scheduling of management activities over a time horizon (Oregon Department of Forestry, 2006).

3.1 General Description

As noted, goal programming models can be devised in a manner very similar to linear or mixed-integer programming. In these models of real-world systems, there exists an objective function, resource and policy constraints, and accounting rows. As in the case of linear programming, where continuous value decision variables are used, the assumptions of additivity, proportionality, certainty, and divisibility still hold (Bare & Anholt, 1976). However, a goal programming model can utilize integer value decision variables just as easily. While heuristic search processes can also be devised with an objective function that minimizes deviations from goals related to disparate forest harvest scheduling outcomes, the following discussion relates more directly to the use of exact mathematical methods for solving problems.

3.2 Objective Function

The objective function of a goal programming model is designed to minimize deviations of resource outcomes from the scheduling of management activities and those target goal achievement levels that are important in the management of a forest. The objective function might be designed in this manner:

$$\text{Minimize} \sum_{t=1}^{T} \sum_{g=1}^{G} \left(w_{t,g}^{-} D_{t,g}^{-} + w_{t,g}^{+} D_{t,g}^{+} \right)$$

where

T represents the number of time periods in the time horizon of the plan

t represents an individual time period

G represents the number of different goals represented in the problem

g represents one goal

$w_{t,g}^{-}$ and $w_{t,g}^{+}$ represent weights for negative and positive deviations from target achievement levels of goals (g), respectively, during each time period (t)

$D_{t,g}^{-}$ and $D_{t,g}^{+}$ represent the under- or overachievement of goals (g), respectively, during each time period (t)

Goal programming is inherently a minimization problem; however, there may be cases where the "goals" are outcomes that a forest manager does not desire, and one might then formulate a problem to maximize the deviations from them. It is not necessary to recognize every combination of g and t in the objective function. Etemad et al. (2019) provided an example where the objective function contained only the negative deviations from goal levels corresponding to total volume scheduled for harvest from a forest, volume by tree species, carbon sequestered, forest growth, labor required, and net present value.

Example

Imagine that a southern United States forest management problem has a time horizon that is 20 years long, and that the time periods are all 1 year long. From accounting rows within the forest management problem, the scheduled timber harvest volume might be contained in variables such as *HV1*, *HV2*, ... *HV20*. From other accounting rows within the forest management problem, the habitat suitability score for the pileated woodpecker (*Dryocopus pileatus*),

(continued)

for example, might be contained in variables such as *HS1*, *HS2*, ... *HS20*. Without interjecting any preference for the goals and their achievement, the objective function would become

$$\text{Minimize } D^-_{HV1} + D^+_{HV1} + D^-_{HV2} + D^+_{HV2} + \ldots + D^-_{HV20} + D^+_{HV20} + D^-_{HS1} + D^+_{HS1} + D^-_{HS2} + D^+_{HS2} + \ldots + D^-_{HS20} + D^+_{HS20}$$

3.3 Goal-Related Constraints

In addition to accounting rows, resource constraints, and policy constraints that are common in linear and mixed integer programming problems, goal-related constraints are included to facilitate minimizing deviations in under- and over-achievement of the desired outcomes. Within the context of the Simplex Method, the deviations (under- or overachievement of goals) are technically slack values represented in two dimensions, the production of some outcome that is either over or under a desired target level (Lee, 1973). The specification of the appropriate target goal levels to achieve through the proposed management of a forest is important, because if the target goal levels are too low, and suboptimal solutions can result (Zeleny, 1981) since overachievement of low goal levels may be penalized.

Imagine for a moment that a forestry organization had a timber harvest production goal in the first time period of a time horizon that was 200,000 tons of wood, and that they also had a habitat suitability goal of 0.5 in the first time period. An accounting row would be devised to summarize the scheduled timber harvest volume, and for the first time period it may be represented with the variable *HV1*. Another accounting row would be devised to summarize the habitat suitability that results after management activities are scheduled, and for the first time period it may be represented with the variable *HS1*.

Conceptually, the goal-related constraints would be

$$HV1 + D^-_{HV1} - D^+_{HV1} = 200000$$

$$HS1 + D^-_{HS1} - D^+_{HS1} = 0.5$$

Using the first of these constraints, perhaps the harvest volume scheduled in the first time period was only 180,000 tons. To then satisfy the first constraint,

$$D^-_{HV1} = 20000$$

$$D^+_{HV1} = 0$$

Using the second of the noted constraints above, perhaps the habitat suitability in the first time period was 0.56. To satisfy the second constraint above,

$$D_{HS1}^{-} = 0$$

$$D_{HS1}^{+} = 0.06$$

With these goal-oriented constraints, one of the two over- or underachievement variables will have a value of 0.0, while the other represents the amount (over or under) or distance from the target goal level. These variables (through their associated constraints) act as a link between accounting rows and the objective function.

3.4 Weights/Normalization

In the allocation of resources or scheduling of activities, goal programming favors higher priority goals, as reflected by their weight in the objective function or their relative magnitude (Hansen, 1977). The larger weights (or preference structures) force the associated deviations toward zero, and thus force the solution to approach the associated target goal levels more rapidly than those outcomes that have smaller weights (Field, 1973). Weighting the achievement of goals effectively ranks the various goals and forces the search process to satisfy as closely as possible the goal with the highest weight (priority) before other less important goals are considered (Bare & Anholt, 1976). Preference structures can be embedded with the weights (Field, 1973), and when a preference structure is imposed, essentially magnifying the weights, a goal programming process further ensures that higher priority deviations from goals are addressed before lower priority deviations. With respect to weights assigned to a common goal, the weighting is said to represent the opportunity cost or *coefficients of regret* to be minimized from not achieving each target goal level (Lee, 1973; Bare & Anholt, 1976).

The importance of the goals, as reflected through their weights, typically allows under- and overachievement, and exact achievement is not required (Sellers & North, 1979). When overachievement is permitted, outcomes greater than the target goal level are acceptable, and when underachievement is permitted, outcomes less than the target goal level are acceptable (North et al., 1976). However, when the weights heavily favor one goal over another, near-exact achievement of the heavily weighted goal might be observed in the solution generated. If overachievement of a goal is acceptable, the variables representing positive deviations from goal levels might be removed from the objective function, whereas if underachievement of a goal is acceptable, the variables representing negative deviations from goal levels might be removed from the problem (Lee, 1973). For example, when only the negative deviations from goals are important (e.g., through concerns about supplying less than a desired level timber to a mill), the variables representing positive deviations from goals might be omitted from a problem (Bare & Anholt, 1976).

Goal weights are often used to normalize the different goals represented in the objective function and penalize in a sense the goal deviations (Sellers & North, 1979). Goal weights can also imply preference or bias, and further, priorities can be imposed through various ordering processes (Diaz-Balteiro et al., 2013). For example, habitat suitability may be measured on a scale of 0 to 1 (or for a stand, or for a forest), while timber production may be measured on a scale of 0 to millions (again or for a stand, or for a forest). Without weighting either of these interests in the objective function, a deviation from a timber production goal would be mathematically greater in magnitude, and thus conceptually more important than a deviation in a habitat suitability score. Assuming in this case that very disparate goals can be normalized, the highest priority goal can be represented mathematically (through further adjustment to the objective function coefficient assigned to the under- or overachievement variables) as the most rigid (Zanakis & Gupta, 1985).

Example

Building on the previous example, perhaps the scheduled harvest volume may range from 0 to 200,000 tons of wood per year, and the habitat suitability score for the entire forest may range from 0 (poor) to 1 (optimal). Therefore, to somewhat normalize the importance of each, perhaps the weights for deviations in scheduled timber volumes are 1.0 in each year, and the weights for deviations in habitat suitability score are 400,000 each year. The objective function would become

$$\begin{aligned}\text{Minimize}\quad & D^-_{HV1} + D^+_{HV1} + D^-_{HV2} + D^+_{HV2} + \ldots + D^-_{HV20} + D^+_{HV20} \\ & + 400000\, D^-_{HS1} + 400000\, D^+_{HS1} + 400000\, D^-_{HS2} + 400000\, D^+_{HS2} + \ldots \\ & + 400000\, D^-_{HS20} + 400000\, D^+_{HS20}\end{aligned}$$

Careful consideration of the objective function coefficients is warranted when attempting to normalize the importance of disparately valued goals. In this case, if the habitat suitability value had been 0.00 in the first time period of the planning problem, the deviation from the goal would have been represented by

$$D^-_{HS1} = 0.5$$

Therefore, in this case, the contribution to the objective function value would have been 400,000 × 0.5, or 200,000. This outcome would represent the same impact on the objective function value had no (zero) timber volume been scheduled during the first time period or had twice as much timber volume been scheduled than desired in the first time period.

If the underachievement of a goal, such as scheduled timber harvest volumes, is deemed more important, the objective function can be modified slightly to indicate this is relatively more important than other concerns.

(continued)

$$\text{Minimize} \quad 2\,D^-_{HV1} + D^+_{HV1} + 2\,D^-_{HV2} + D^+_{HV2} + \ldots + 2\,D^-_{HV20} + D^+_{HV20} \\ + 400000\,D^-_{HS1} + 400000\,D^+_{HS1} + 400000\,D^-_{HS2} + 400000\,D^+_{HS2} + \ldots \\ + 400000\,D^-_{HS20} + 400000\,D^+_{HS20}$$

Here, the coefficient "2" was placed in front of the decision variables representing underachievement of timber volume to imply that under-producing timber harvest volumes is twice as important as over-producing timber harvest volumes. Certainly, any value other than "1" (the default coefficient that is not necessary to show in the objective function) would imply greater preference. By increasing the coefficient values related to the underachievement of timber harvest volumes, the intent is to also indicate it is more important than the achievement of habitat suitability goals. If the outcomes from the model suggest otherwise, perhaps a coefficient of greater magnitude is needed to emphasize the notion that an underachievement of the timber harvest goal is indeed more important than deviations from the habitat suitability goals.

4 Contemporary Forest Harvest Scheduling Problem

The small forest management problem from Sect. 4 of Chap. 2 included a single objective: maximize discounted net revenue from final harvests scheduled to each timber stand within the time horizon of the problem. With goal programming, the objective function would contain two or more interests, and goal targets for these would be proposed. Deviations from the goal targets would then be minimized within the objective function. Therefore, within an accounting row, the discounted net revenue could be accumulated. To accomplish this, the relationships in the objective function from the problem in Sect. 4 of Chap. 2 would be moved to an accounting row which would be further enhanced to provide the total discounted net revenue through a single variable (actual discounted net revenue, or Actual_DNR below).

```
318808.87 S1P1 + 313359.15 S1P2 + 306872.63 S1P3 + 299955.48 S1P4 + 292070.00 S1P5 +
283779.09 S1P6 + 275332.00 S1P7 + 266459.46 S1P8 + 257404.97 S1P9 + 248408.31 S1P10 +
239315.76 S1P11 + 230219.04 S1P12 + 221144.39 S1P13 + 212253.69 S1P14 + 203561.19 S1P15 +
...
0.00 S31P1 + 0.00 S31P2 + 0.00 S31P3 + 0.00 S31P4 + 0.00 S31P5 + 0.00 S31P6 + 0.00 S31P7 +
0.00 S31P8 + 46793.68 S31P9 + 48989.33 S31P10 + 50529.01 S31P11 + 51555.26 S31P12 +
52131.57 S31P13 + 52369.37 S31P14 + 52260.26 S31P15- Actual_DNR = 0
```

Further, an equation would be developed to determine how far the actual discounted net revenue lies above or below the goal target assumed by the forest planner.

```
Actual_DNR - DRN_positive_deviation + DRN_negative_deviation =
DNR_goal_target
```

For example, if the goal target was assumed to be $3,500,000 (US dollars) and the actual discounted net revenue was $3,597,228, the positive deviation from the goal target would be $97,228, and the negative deviation would be $0.

If another goal of interest concerned wood production, the outcomes of scheduled harvest volumes could be quantified as a response to the assigned management activities. In the previous problem, scheduled harvest volumes were accumulated for each time period through distinct variables (VC1, VC2, etc.). Here, different goal targets for each time period could be proposed by the forest planner, and deviations from these could be determined.

```
VC1 - VC1_positive_deviation + VC1_negative_deviation =
VC1_goal_target
...
VC15 - VC15_positive_deviation + VC15_negative_deviation =
VC15_goal_target
```

Finally, the objective function for the goal programming problem would be designed to minimize the deviations from the various goal targets.

```
Minimize
DRN_positive_deviation + DRN_negative_deviation +
VC1_positive_deviation + VC1_negative_deviation + ... +
VC15_positive_deviation + VC15_negative_deviation
```

Of course, since the magnitude of the deviations goal targets is different, assuming each outcome is equally important weights might be associated with each variable in the objective function to normalize the outcomes. For example, the deviations related to the discounted net revenue goal target may range from zero to hundreds of thousands of dollars, while the deviations related to the scheduled harvest volumes may range from zero to thousands of tons of wood. The volume-related variables might, in this case, be multiplied by a weight greater than 1.0 to ensure that the deviations from volume-related goal targets are just as important as the deviation from the discounted net revenue goal target in the solution to the forest harvest scheduling problem.

In an example where the goal target for the discounted net revenue is $3,500,000, the goal target is 20,000 tons of scheduled harvest volume in each of the fifteen time periods, and fractional harvest decisions can be assigned to timber stands, the following abbreviated results were produced:

```
Global optimal solution found.
Objective value:        8066.617
Total solver iterations:     306
Elapsed runtime seconds:    0.11

Variable                Value
DNR_positive_deviation     0.000
DNR_negative deviation     0.000
VC1_positive_deviation     0.000
VC1_negative deviation  8066.617
VC2_positive_deviation     0.000
VC2_negative deviation     0.000
VC3_positive_deviation     0.000
VC3_negative deviation     0.000
VC4_positive_deviation     0.000
...
ACTUAL_DNR    3500000
...
VC1        11933.38
VC2        20000.00
VC3        20000.00
VC4        20000.00
VC5        20000.00
VC6        20000.00
VC7        20000.00
VC8        20000.00
VC9        20000.00
VC10        20000.00
VC11        20000.00
VC12        20000.00
VC13        20000.00
VC14        20000.00
VC15        20000.00
```

In this solution to the forest harvest scheduling problem, the goal target for net discounted revenue ($3,500,000) is achieved. Furthermore, the goal targets for the desired harvest levels (20,000 tons per year) are achieved, with the exception of time period 1 where the scheduled volume (11,933 tons) has a negative deviation from the target of 8067 tons. It seems from this problem formulation that the ability to achieve a $3,500,000 net discounted revenue target and the ability to produce 20,000 tons of wood per year is not possible. Should one decide that a negative deviation in the scheduled wood volume during time period 1 is more important than missing the goal target in other time periods, a weight could be applied in the objective function to indicate relative importance of this desire.

```
Minimize
DRN_positive_deviation + DRN_negative_deviation +
VC1_positive_deviation + 5 VC1_negative_deviation + ... +
VC15_positive_deviation + VC15_negative_deviation
```

The outcome associated with this revised harvest scheduling problem is interesting. Given the new formulation, it seems that the shortfall in scheduled harvest volume has been transferred to time period 2.

```
Global optimal solution found.
Objective value:        8512.250
Total solver iterations:    305
Elapsed runtime seconds:    0.13

Variable       Value
DNR_positive_deviation     0.000
DNR_negative deviation     0.000
VC1_positive_deviation     0.000
VC1_negative deviation     0.000
VC2_positive_deviation     0.000
VC2_negative deviation  8512.250
VC3_positive_deviation     0.000
VC3_negative deviation     0.000
VC4_positive_deviation     0.000
...
ACTUAL_DNR    3500000
...
VC1         20000.00
VC2         11487.75
VC3         20000.00
VC4         20000.00
VC5         20000.00
VC6         20000.00
VC7         20000.00
VC8         20000.00
VC9         20000.00
VC10        20000.00
VC11        20000.00
```

```
VC12        20000.00
VC13        20000.00
VC14        20000.00
VC15        20000.00
```

5 Applications

Hotvedt (1983) described the application of goal programming to a harvest scheduling problem for an 84,735 acre (34,292 ha) forest in the southern United States. The time horizon of this problem was 90 years, which was divided into 5-year time periods. The forest was nearly equally composed of pine (*Pinus* spp.) plantations and naturally regenerated pine-hardwood (e.g., *Quercus* spp.) forests. After final harvests were scheduled, it was assumed that amenable areas were to become pine plantations. One set of constraints in the planning problem limited the amount of management activity within each forest age class in an attempt to direct the forest toward a regulated state. Other constraints were related to economic returns, and the desire to ensure desirable periodic cash flows. Goals were designed for periodic schedule harvest levels, periodic cash flows, and periodic costs of forest management. Example solutions illustrated the outcomes when preferences for minimizing deviations from various sets of goals were acknowledged.

As a second example, Demirci and Bettinger (2015) demonstrated the application of goal programming to an interesting forest harvest scheduling problem in the Republic of Turkey. The management problem concerned a 6464 acre (2616 ha) forest in western Turkey that consisted of Anatolian black pine (*Pinus nigra*) and oaks (*Quercus* spp.). This area was divided into 106 compartments. Harvest decisions were assigned over a 20-year time horizon, which was divided into 1-year time periods. In the course of scheduling thinning activities for pine forests, it was suggested that two issues were important: (1) volumes scheduled for harvest each year needed to be similar, and (2) management units where thinnings were scheduled needed to be not too far apart (to keep thinning crews in the same general area each year). Therefore, the objective function of the goal programming problem contained variables that represented deviations in scheduled harvest volumes from an assumed target volume, and variables that represented deviations in distances between the centers of each scheduled thinning unit from some desired distance. In minimizing both concerns, a tactical plan was developed that included relatively even wood flows from final harvests and thinning activities, and a relatively clustered spatial arrangement of thinning activities within a forest.

As a third example of the use of goal programming for harvest scheduling, the problem centered on a 367-acre (149 ha) forest of *Eucalyptus globulus* plantations in Spain, where the overall strategy was to arrive at a regulated forest structure at the end of one rotation (Bertomeu et al., 2009). The goals within the objective function were designed to obtain a relatively even amount of land area within each age class at the end of the planning horizon, to schedule a relatively even amount volume for harvest throughout the planning horizon, and to limit deviations from a target net present value for the forest as a whole. In addition to resource constraints, limits on the minimum age at which a stand of eucalypts can be harvested and limits on the minimum size of harvest area (to ensure profitability) were imposed.

As a final example of the use of goal programming in forestry, a very large set of goals were devised to assist in the development of a forest management plan. The Hiawatha National Forest encompasses nearly one million acres (about 400,000 ha) in two large blocks of land (Fig. 4.1) located on the Upper Peninsula of the State of Michigan (United States). The national forest was created in 1931 from lands abandoned by farmers, lands affected by fires, and lands where trees were harvested and left to naturally regenerate (US Department of Agriculture, Forest Service, 2006). In the development of the 2006 forest plan, one alternative assessed by the planning team had the purpose of transitioning the forest structure from its current state to a desired future structural state as soon as possible. In this analysis, about 120 goals were devised, many of which described as minimum and maximum amounts of various vegetation types (desired future conditions) to be developed over time. The objective function was subsequently designed to minimize under- or overachievement of the large set of goals (US Department of Agriculture, Forest Service, 2006).

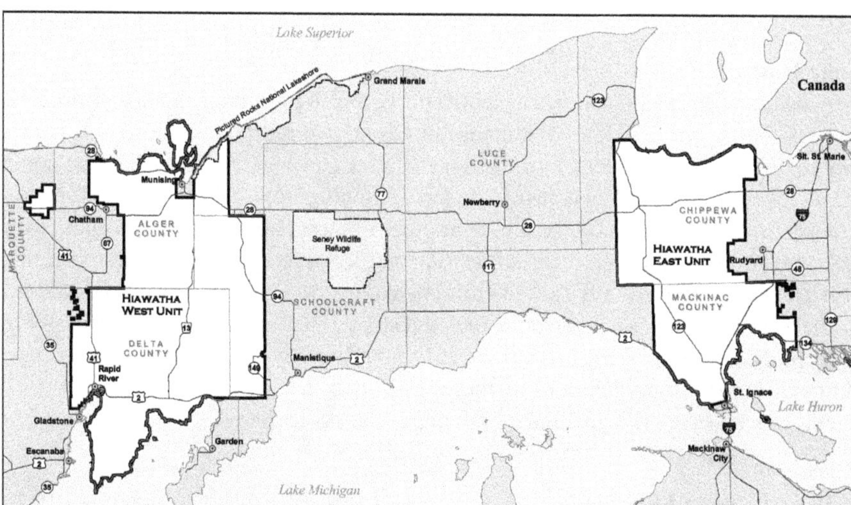

Fig. 4.1 Location of the Hiawatha National Forest in Michigan, USA. Credit: US Department of Agriculture, Forest Service, 2006

6 Summary

Goal programming may, in some circumstances, more closely emulate the logic employed by forest managers when multiple-use forest management is practiced (Schuler et al., 1977). Goal programming has been characterized as a flexible analytical framework for including diverse interests in the objective function of complex forest harvest scheduling problems, and as a process that then satisfices different interests that concern the management of forests (Diaz-Balteiro et al., 2013). Some have suggested that the *highest and best use* of resources within a landscape might be determined using this type of process, even though conflicts among goals and limitations within resources may be evident (North et al., 1976). Solving a forest harvest scheduling problem in this manner can also illustrate which goals cannot be achieved under different management alternatives (Lee, 1973). Interestingly, whether the Simplex Method is employed for linear goal programming problems or the branch and bound (or cutting plane) is employed for mixed integer goal programming problems, an optimal solution to the specific alternative under consideration (as reflected through the weights) may be located, yet it could be neither economically nor ecologically efficient for the model system under consideration, as the goal weights and goal deviations may only produce a satisfactory solution due to the mixture of desired outcomes (North et al., 1976; Sellers & North, 1979).

Take Home Message

In general, when using an exact algorithm, goal programming is an optimization process that is very similar in structure to linear programming and mixed integer programming. The problem formulation includes an objective function, and some resource constraints, policy constraints, and accounting rows. The objective function is different from linear and mixed integer programming; however, in that, more than one type of resource (or measurable outcome) is recognized. Furthermore, the objective function generally seeks to minimize deviations from defined goals related to the outcomes acknowledged in the objective function. As a problem-solving concept, goal programming can also be accommodated by a heuristic search process, although the resulting solutions in these cases may not be optimal since heuristic search processes cannot guarantee optimality will be achieved.

Questions

1. If the volume to be produced by two very disparate resource outcomes (e.g., blueberries and timber) were described as goals to achieve in a goal programming problem, how might you adjust a problem formulation so that they have the same emphasis or influence on the resulting solution (forest harvest schedule)?
2. One advantage of goal programming over other related problem-solving methods is that a feasible solution to a problem (aside from problems caused by typographical errors in the problem formulation) might always be produced. Why?

3. In your opinion, would the inclusion of an issue of concern (e.g., the development of wildlife habitat) in the objective function of a goal programming problem formulation, rather than a recognition of the achievement of this issue of concern through the use of constraints, make this concern more important for the people who implement the resulting harvest schedule?
4. In general, how might one develop the goal targets for goal programming problem that is designed to produce a forest harvest schedule?
5. If one of the goals in a goal programming problem involves the production of timber from a forest, would the positive or negative deviations from a goal be more important to you? Why?

References

Bare, B. B., & Anholt, B. F. (1976). *Selecting forest residue treatment alternatives using goal programming*. (General Technical Report PNW-43). U.S. Department of Agriculture, Forest Service, Pacific Northwest Forest and Range Experiment Station.

Bell, E. F. (1977). Mathematical programing in forestry. *Journal of Forestry, 75*(6), 317–319.

Bertomeu, M., Diaz-Balteiro, L., & Gimenez, J. C. (2009). Forest management optimization in *eucalyptus* plantations: A goal programming approach. *Canadian Journal of Forest Research, 39*(2), 356–366.

Charnes, A., & Cooper, W. W. (1961). *Management models and industrial applications of linear programming* (Vol. volume I). John Wiley & Sons, Inc.

Charnes, A., Haynes, K. E., Hazleton, J. E., & Ryan, M. J. (1975). A hierarchical goal-programming approach to environmental land use management. *Geographical Analysis, 7*(2), 121–130.

Demirci, M., & Bettinger, P. (2015). Using mixed integer multi-objective goal programming for stand tending block designation: A case study from Turkey. *Forest Policy and Economics, 55*, 28–36.

Diaz-Balteiro, L., González-Pachón, J., & Romero, C. (2013). Goal programming in forest management: Customising models for the decision-maker's preferences. *Scandinavian Journal of Forest Research, 28*(2), 166–173.

Etemad, S. S., Limaei, S. M., Olsson, L., & Yousefpour, R. (2019). Forest management decision-making using goal programming and fuzzy analytic hierarchy process approaches (case study: Hyrcanian forests of Iran). *Journal of Forest Science, 65*(9), 368–379.

Field, D. B. (1973). Goal programming for forest management. *Forest Science, 19*(2), 125–135.

Hansen, B. G. (1977). *Goal programming: A new tool for the Christmas tree industry*. (Research Paper NE-378). U.S. Department of Agriculture, Forest Service, Northeastern Forest Experiment Station.

Hotvedt, J. E. (1983). Application of linear goal programming to forest harvest scheduling. *Southern Journal of Agricultural Economics, 15*(1), 103–108.

Killen, J. (2021). *Mathematical programming methods for geographers and planners*. Routledge.

Lee, S. M. (1973). Goal programming for decision analysis of multiple objectives. *Sloan Management Review, 14*(2), 11–24.

North, R. M., Neely, W. P., & Carlton, R. L., Jr. (1976). *The highest and best uses of the Ocklawaha River basin and Lake Rousseau for the economy and the environment*. U.S. Department of the Army, Corps of Engineers.

Oregon Department of Forestry. (2006). *Harvest & Habitat Model project, final report*. Oregon Department of Forestry. Retrieved October 10, 2024, from https://www.oregon.gov/ODF/Documents/WorkingForests/HarvestHabitatModelProject.pdf

Schuler, A. T., Webster, H. H., & Meadows, J. C. (1977). Goal programing in forest management. *Journal of Forestry, 75*(6), 320–324.

Sellers, J., & North, R. M. (1979). A viable methodology to implement the principles and standards. *Water Resources Bulletin, 15*(1), 167–181.

U.S. Department of Agriculture, Forest Service. (1984a). *Draft environmental impact statement for the Caribou National Forest proposed land & resource management plan*. U.S. Department of Agriculture, Forest Service, Intermountain Region.

U.S. Department of Agriculture, Forest Service. (1984b). *Draft environmental impact statement, Kisatchie National Forest land and resource management plan*. U.S. Department of Agriculture, Forest Service, Southern Region.

U.S. Department of Agriculture, Forest Service. (2006). *Hiawatha National Forest final environmental impact statement to accompany the 2006 land and resource management plan*. U.-S. Department of Agriculture, Forest Service, Eastern Region.

Zanakis, S. H., & Gupta, S. K. (1985). A categorized bibliographic survey of goal programming. *OMEGA International Journal of Management Science, 13*(3), 211–222.

Zeleny, M. (1981). Technical note: The pros and cons of goal programming. *Computers & Operations Research, 8*(4), 357–359.

Chapter 5
Binary Search

Learning Objectives
The objectives of this chapter on binary search are to help the reader

- Understand the history and motivation of the binary search process
- Understand the general manner in which binary search operates
- Understand the use and value of sequential scheduling and how it might help a search process
- Understand the main limitations associated with the use of binary search

1 Introduction

Binary search is a simulation process that may provide feasible solutions to certain forest planning problems that have a desired target of produceable outcomes for each time period of the planning horizon. Under more general circumstances, binary search has been noted as one of the most classical algorithms described in the published literature (Lesuisse, 1983). However, optimality is not guaranteed with this search process. In conducting a binary search simulation, upper and lower bounds for the desired produceable outcomes are identified, and the midpoint of these is used as the target for a subsequent attempt to develop a feasible harvest schedule. These bounds, and the target, are often defined in forestry as potential harvest levels desired during a specific period of time. The range of feasible solutions (defined as the distance between the upper and lower bounds on the objective function value) contracts until the gap between them is so small that it is reasonable to stop searching for a better solution (Hann & Brodie, 1980). The simulation process works well when there is a commodity production objective, but any goal that reflects something being produced from a forest (e.g., revenue, habitat) can be accommodated.

P. Bettinger, *Forest Harvest Scheduling*, Springer Texts in Business and Economics,
https://doi.org/10.1007/978-3-031-89432-9_5

2 Background

Binary search, *half-interval search, dichotomizing search,* or *search by partition* (Halpern, 1958; Bureau of the Budget, 1962), is a simulation process for locating a very good solution to a problem by progressively cutting in half a range (defined by upper and lower bounds) of potential produceable outcomes (Davis, 1969). Binary search has also been characterized as a *re-start procedure to solve convergence problems* (Lyon & Sedjo, 1986) and a *heuristic method* (Sessions et al., 2000). Paraphrasing Seidl and Enderle (2011), a very technical description of binary search involves:

- Starting with the middle element in a sorted list of elements bounded on both ends with known values.
- Comparing the middle element to a search key.
- Terminating the search if the middle element is the search key.
- If termination does not occur, the middle element observed becomes one of the two bounds of a new subset of the sorted list of elements, and a new comparison is made (the process then returns to the second bullet point above).

Binary search is a dichotomizing search where a sorted list of values is divided into two approximately equal subsets; one desirable subset is retained while the other undesirable subset is discarded (Davis, 1969). In a more general sense, binary search is a problem-solving procedure where a series of items in a sorted list is divided into two parts during one iteration of the process, one of which is deemed acceptable and the other deemed unacceptable. Within the acceptable part, the search continues in a similar manner until the desirable property is located (Bureau of the Budget, 1962). In addition to problems designed to efficiently locate text strings from a sorted list (Davis, 1969), early applications of binary search focused on a number of engineering problems (Katz et al., 1961).

Example

As a simple example of binary search, imagine that Person A says to Person B, *"Select a number between 1 and 1000 and do not reveal it to me. I can locate your secret number in less than 11 guesses if you will only tell me whether my guess is higher or lower than your secret number."* Further imagine that Person B selects 256 as their number. At the onset of the search for Person B's number, Person A only knows that Person B's number lies between 1 and 1000. And there are 1000 potential answers to the problem. Shown below is a binary search approach Person A might employ to locate Person B's mysterious number in less than 11 attempts.

Upper bound = 1000
Lower bound = 1

(continued)

Attempt #1: Person A suggests that Person B's number is 500 (midway between the upper and lower bound).

Person B replies that this guess is too high.

New upper bound = 500
Lower bound = 1

Attempt #2: Person A suggests that Person B's number is 250 (midway between the upper and lower bound).

Person B replies that this guess is too low.

Upper bound = 500
New lower bound = 250

Attempt #3: Person A suggests that Person B's number is 375 (midway between the upper and lower bound).

Person B replies that this guess is too high.

New upper bound = 375
Lower bound = 250

Attempt #4: Person A suggests that Person B's number is 312 (about midway between the upper and lower bound).

Person B replies that this guess is again too high.

New upper bound = 312
Lower bound = 250

Attempt #5: Person A suggests that Person B's number is 281 (midway between the upper and lower bound).

Person B replies that this guess is again too high.

New upper bound = 281
Lower bound = 250

Attempt #6: Person A suggests that Person B's number is 265 (about midway between the upper and lower bound).

Person B replies that this guess is again too high.

New upper bound = 265
Lower bound = 250

Notice at this point, after only 6 guesses, that the range of values of Person B's secret number is 250 to 265. Certainly, with only 16 more guesses (the number of options between 250 and 265), Person A can locate Person B's number. But let's continue with binary search.

Attempt #7: Person A suggests that Person B's number is 257 (about midway between the upper and lower bound).

(continued)

Person B replies that this guess is again too high.

New upper bound = 257
Lower bound = 250

Attempt #8: Person A suggests that Person B's number is 253 (about midway between the upper and lower bound).
Person B replies that this guess is too low.

Upper bound = 257
New lower bound = 253

Attempt #9: Person A suggests that Person B's number is 255 (midway between the upper and lower bound).
Person B replies that this guess is again too low.

Upper bound = 257
New lower bound = 255

The only other option at this point is to guess that Person B's number is 256, which is located on the tenth guess of Person A, perhaps to the astonishment of Person B.

In forestry, one might use binary search to develop a forest plan that provides the highest and most even wood flow across a time horizon. Binary search processes inherently have only two choices once harvests have been simulated across a time horizon: increase the proposed harvest level or decrease the proposed harvest level (Johnson & Tedder, 1983). The amount of the periodic wood flow would represent the targeted values, and perhaps it initially has a wide range (e.g., 0–1,000,000 tons per time period). So, imagine that the sorted list is the range of potential scheduled wood flow values (e.g., 0–1,000,000 tons per time period). The middle value (500,000 tons) is selected and tested for feasibility, both in meeting this level of wood flow in each time period of the time horizon and in satisfying all other constraints. Once feasibility is assessed (yes or no), either the range from 0 to 500,000 tons is retained (if scheduling 500,000 tons per time period is not feasible), or the range 500,000 to 1,000,000 tons is retained (if scheduling 500,000 tons is feasible). The retained subset of then subjected to further similar division if some pre-defined stopping criterion has not been met. These types of search processes represent some of the simplest and most efficient search methods (Manna & Waldinger, 1987).

If it was our desire to locate the best forest harvest schedule (as measured by scheduled wood flow) with proposed harvest levels between 0 and 1,000,000 tons per time period, we could assess every possible periodic harvest level through complete enumeration of the possibilities. In this case, we might have to assess 1,000,001 potential solutions, if discrete, integer representations of scheduled tonnage were assumed (e.g., no tons per year, 1 ton per year, 2 tons per year, 3 tons per year, etc.). Yet, with binary search, the number of potential harvest levels to assess would be represented by the value of the truncated solution to

$$1 + \mathrm{Log}_2(N)$$

where N is the total number of potential harvest levels (1,000,001). And so, with binary search we would only need to assess

$$1 + \mathrm{Log}_2(1{,}000{,}001) = 20.93 \text{ or } 20 \text{ potential periodic wood flow levels}$$

If one were faced with the choice of assessing 1,000,001 or 20 potential wood flow levels, 20 would obviously be completed much more efficiently.

In the example provided earlier, it was suggested that Person A could locate Person B's mysterious number between 1 and 1000 in less than 11 attempts or guesses. In that case, $1 + \mathrm{Log}_2(1000) = 10.97$ potential guesses to locate the number. Thus, it is obvious that as the number of values in a list (or nodes in a search tree) increases, the maximum number of potential guesses during a binary search process increases (Overholt, 1973).

3 Process

As was noted earlier, binary search is a simulation method for solving problems. During a single time period, two general issues are considered in a binary search process: the amount of wood volume to schedule for harvest, and the order of harvest from various stands or strata (Schmidt & Tedder, 1981). Binary search has generally been applied to harvest scheduling problems that seek to locate the highest, and most even feasible harvest levels over some period of time. This suggests that a single harvest level *target* is employed in order to guide the search, although others have noted that a *pattern* of harvests (not necessarily a single desired harvest level each time period) that relaxes the even-flow nature inherent of this process can also be employed, to schedule harvests that best emulate the pattern (Bettinger et al., 2005). The target harvest level for a single iteration of the model represents one potential end (or bracket) of a range that defines the upper or lower bounds. In other words, the outcome (success or failure in scheduling the target harvest level) becomes either an upper or lower bound once the result is known. These bounds or adaptive borders of the search (Seidl & Enderle, 2011) change with each iteration of the model.

3.1 General Description

In developing a forest harvest schedule, the basic flow (Fig. 5.1) of a binary search process that seeks to maximize a wood flow objective may involve the following:

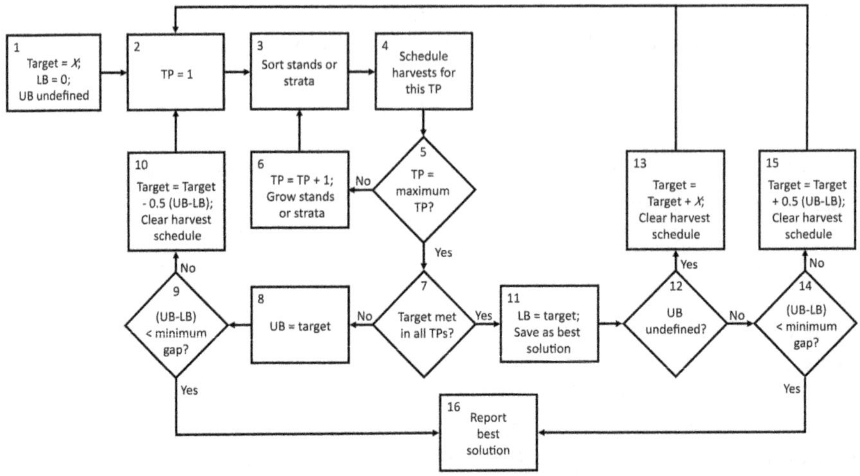

Fig. 5.1 A binary search process for forest harvest scheduling, where TP = a single time period; Target = the desired harvest level each time period; LB = lower bound on the feasible harvest level; UB = upper bound on the harvest level (yet infeasible); minimum gap = the stopping rule where the change in proposed harvest level is so small that it is reasonable to end the simulation; and X = the initial harvest level target assumed by the forest planner. Credit: Pete Bettinger

- Define the initial target harvest level (Step 1), and remember it as X.
- For a single time period, beginning with time period 1 and progressing to time period n …
  - Sort the potential harvest areas using a pre-defined metric as the sorting key (Step 3).
  - Schedule harvests up to, or perhaps slightly past X (in the case of integer decisions applied to stands), the harvest target (Step 4).
  - Once Step 4 is complete for a time period, grow the forests and schedule harvests for the next time period (Step 6).
  - Once all time periods have harvests scheduled, assess whether it was possible to meet the harvest target (X) in all time periods (Step 7).

    If the harvest target could not be obtained in any one time period, X becomes an upper bound (Step 8), where it is clear that the harvest target is not obtainable. From here, one would assess whether it is time to terminate the simulation (Step 9) or continue (try again) with a new harvest target (X') that is midway between the upper and lower bounds (Step 10).

    If the harvest target could be obtained in all time periods, X becomes a lower bound (Step 11) if the solution is feasible. From here, if the upper bound is still undefined, X is increased (Step 13) and the simulation continues (tries again) with a new harvest target. If the upper bound has previously been defined, one would assess whether it is time to terminate the simulation (Step 14) or continue (try again) with a new harvest target (X'') that is midway between the upper and new lower bound (Step 15).

Other than the first few iterations of the model, where either the upper or lower bounds are not yet evident, success (lower bound) or failure (upper bound) is learned through the attempt to schedule a certain level of harvest over time. The upper and lower bounds will not both be known until the first failure to schedule a harvest level is encountered (Steps 7–8). The failed harvest level becomes the upper bound and the last successfully scheduled harvest level becomes the lower bound. The very next attempt (harvest target) is midway between the two harvest levels, and

- If the next proposed harvest target is unobtainable in any one time period, the proposed harvest target becomes the new upper bound for the next iteration of the model.
- If the next proposed harvest target is obtainable in every time period, the proposed harvest target becomes the new lower bound for the next iteration of the model.

In describing the ECHO binary search model, Walker (1975) suggested the following for a strata-based planning process, where age classes of forests (not individual stands) are scheduled for harvest:

> *The ... harvest level is ... adjusted up or down by some specified amount and the whole calculation procedure starts again with this new harvest level. ... The next trial harvest level is then set midway between these two levels and the iterations continue, always setting the successive trial harvest levels midway between the last two... The calculations are stopped when the difference between the two levels ... becomes sufficiently small that continued iterations are unwarranted. This process is therefore called an iterative optimum-seeking binary search technique.*

In essence, binary search attempts to schedule harvests in such a manner that when target harvest levels are achieved, and all constraints are satisfied, the resulting forest harvest schedule is feasible. Within the simulation process, harvests can be scheduled from strata of forest classes, or from individual stands, either of which are sorted based on some criteria that relates to the timing of harvest decisions. Both a minimum harvest age and a minimum economic achievement criterion can be employed to limit the scheduling of harvests to strata or stands. Further, should individual stands be employed as the simulation unit, harvest adjacency and green-up rules can be employed to limit the timing and placement of harvest units (Bettinger et al., 2005).

3.2 Stand or Strata Sorting Rule

At times, binary search processes utilize fairly large strata (collections of individual stands that have similar characteristics) (Brodie & Sessions, 1991); however, binary search can also be employed at the forest level using individual stand-level decisions. In either case, binary search schedules activities from a sorted list of potential harvest decisions. The sorting rule for forest harvesting decisions often relates to the age class of the strata or the age of each individual stand of trees (Step 3 of the process noted above). However, a more sophisticated sorting rule might be employed to determine which strata or stand should be scheduled for harvest before all other strata or stands. The sorted list therefore provides an established priority of harvest (Schmidt & Tedder, 1981).

Assume here that the sorted list contains even-aged timber stands or strata that are ready for a final harvest and thus contain merchantable wood. The sorting process would arrange them in order, from the highest valued to the lowest valued. The value used in the sorting process is a decision of the planner. The list of strata or stands, for example, could be sorted by age (oldest first), volume (total available for harvest), volume per unit area harvestable, value, or some other metric that facilitates sorting. For example, value ratio (value divided by age) has been used to develop a sorted list for a binary search process (Bettinger & Johnson, 2003). Toward the bottom of the sorted list are those strata or stands with the lowest values. One would need to decide whether to include in the sorted list those strata or stands that are unavailable for harvest because (a) they are too young, or particularly for stands, (b) they recently incurred a management treatment (e.g., thinning, fertilization) that requires waiting a certain period of time before a final harvest. These strata or stands would not be totally forgotten during the binary search process and will appear (or re-appear in later time periods) when conditions are suitable for a final harvest.

Importantly, forestry problems include a time horizon and a set of individual time periods. When applied to forestry problems, binary search attempts to schedule harvest activities in sequential order of the time periods. In other words, potential harvest activities are scheduled for time period 1 before similar activities are scheduled for time period 2 (and time period 2 before time period 3, and so on). The sorting of the list of potential harvest areas is conducted prior to the scheduling of the first harvest activity. Potential harvests are then selected from the top of the list. However, some of the strata or stands could be passed over (skipped) during the scheduling process as a result of constraint violations. For example, within a specific time period, binary search attempts to schedule the first available stratum or stand at the top of the sorted list. If, in attempting to do so, one or more constraints are violated, the stratum or stand may be passed over for harvest, and the next available stratum or stand in the sorted list would be considered. The passed over (skipped) strata or stands would be available for harvest in the next time period, and likely will be found at or near the top of the sorted list. However, some harvest opportunities may be passed over several times before their presence in a harvest schedule does not lead to a constraint violation.

Should the sorted list contain (a) strata or (b) stands that do not need to be completely harvested in one time period, the portion of these strata or stands that is necessary to meet the harvest target might be scheduled for harvest. The remaining portion is then available for harvest in the subsequent time period. This emulates the divisibility principle of linear programming. However, integer assignments may need to be employed to ensure scheduling harvest activities to entire strata or stands.

Accounting for forest thinning opportunities is challenging within binary search, as the most common activity associated with the sorted list is a final harvest. In these cases, the thinnings likely might be scheduled first, subject to some rules (oldest stand ready for thinning first, thinning timing associated with the best stand-level management regime, thin 10% of all stands in each time period, etc.). Then, the remaining volume needed to meet the current volume target of the binary search process might be obtained from strata or stands in the sorted list of potential final harvest activities.

3.3 Harvest Target

When applied to forest management problems, the target value used in binary search is typically some sort of measure of wood flow. These future, scheduled harvest volumes or weights of forest products could be expressed in tons, m^3, MBF (thousand board feet), or other locally accepted metric as long as it can be estimated based on the growth and management history of each stand or strata within a forest. One assumption with regard to the target harvest value in binary search may be that it is constant across time periods of the time horizon. Since binary search schedules harvests one period at a time, the harvests would be scheduled until the target harvest volume (or weight) has been reached. This would begin with time period 1 and proceed to each subsequent time period in order until the end of the time horizon. If the target harvest volume (or weight) was possible to obtain in all time periods within a time horizon, and assuming constraints have not been violated, the target harvest level is feasible. This amount then represents a new lower bound on the maximum sustainable harvest level. Once the scheduling process is unable to produce the desired volume or weight target in one or more time periods, the attempted target becomes an upper bound on the maximum harvest level (Fig. 5.2). At this point in the binary search process, we know the following:

Upper bound = last infeasible harvest level
Lower bound = last feasible harvest level

This is where the half-interval search partitioning begins. As has been noted earlier, the next iteration of the binary search process attempts to produce a harvest level representing the mid-point between the upper and lower bounds (Fig. 5.3). The result of this attempt to schedule harvests to meet the desired target will replace either the upper or lower bound. With each passing iteration, the range between the upper and lower bounds is divided in half. When the range between the upper and lower bounds has been reduced to some trivial amount, the search process terminates, and the last feasible harvest schedule is reported. When there are an odd number of harvest level options, including and between the upper and lower bounds, the mid-point value will be obvious. When there are an even number of harvest level

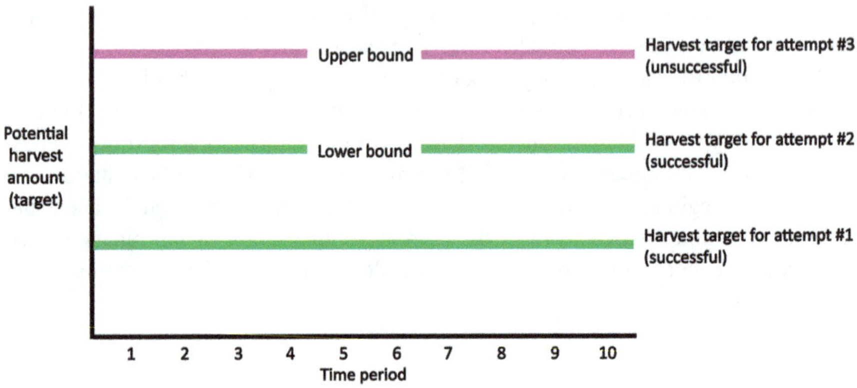

Fig. 5.2 Three different harvest level targets, two successfully met and one that was not successfully met, where the upper and lower bounds on scheduled timber harvests are identified. Credit: Pete Bettinger

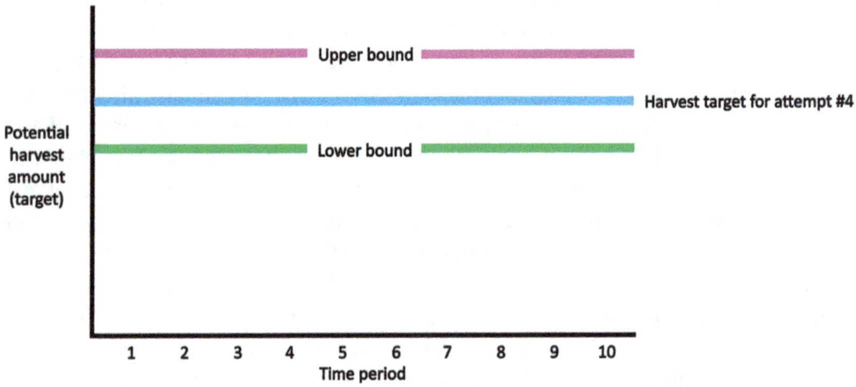

Fig. 5.3 Harvest level targets, where the fourth attempt represents the mid-point between the upper and lower bounds. Credit: Pete Bettinger

options, between the upper and lower bounds, the mid-point value might be selected based on an assumed rule (i.e., choose the lower value of the two at the mid-point, or choose the higher value) or by adding an artificial value to one end of the bounds to ensure that there are an odd number of options including and between the upper and lower bounds (Lesuisse, 1983).

An alternative assumption regarding the target harvest volume is to attempt to produce the highest *pattern of harvest* over time. Rather than a single constant target harvest value, a different amount could be desired during each time period of the time horizon, and this collection represents a pattern of desired harvest volumes. Then, with each iteration of the model, the entire pattern would either be shifted upward or downward, depending on the result of the attempt to schedule harvests to obtain the pattern (Fig. 5.4). A third alternative assumption regarding the target

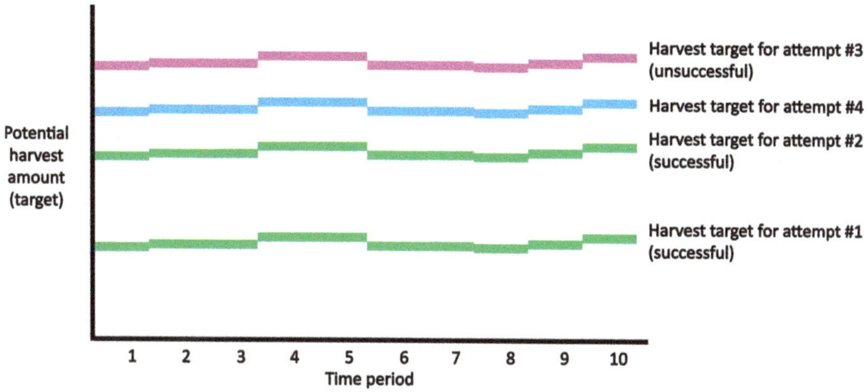

Fig. 5.4 Harvest targets in a case where a pattern of harvest is desired. Credit: Pete Bettinger

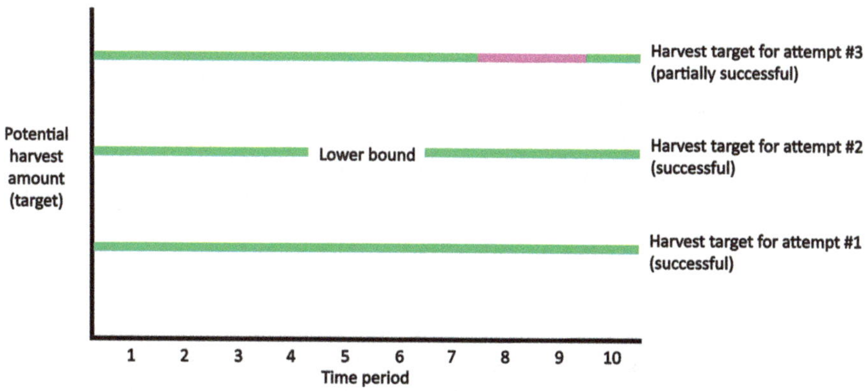

Fig. 5.5 Harvest target missed in time periods 8 and 9 during attempt #3. Credit: Pete Bettinger

harvest volume would involve changing the harvest levels only for time periods where the target volume was unable to be obtained. In these cases, the target harvest level for all feasible time periods would be held constant (fixed), while a set set of target harvest levels for the infeasible time periods would be adjusted (Fig. 5.5). In essence, from this point, the binary search process would be applied just to the target harvest levels of the time periods where the scheduled harvest levels were initially unobtainable (Fig. 5.6).

3.4 Harvest-Level Increment and Stopping Rule

Once harvest activities have been scheduled in each of the time periods, from the first to the last in sequential order, the next step in the binary search process is to assess success (Lesuisse, 1983). For a forestry problem, the assessment involves

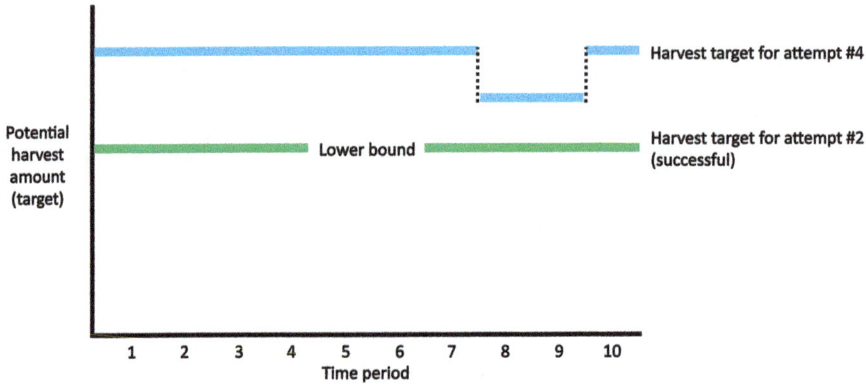

Fig. 5.6 Adjustment of harvest targets for attempt #4 only during time periods where previously scheduled harvest levels were below the target level. Credit: Pete Bettinger

determining whether the harvest target was obtained in each time period. If the determination is positive, the forest harvest schedule is saved if it represents the best solution found so far. If the stopping criterion has not been met, the harvest target increases. If the determination is negative, and if the stopping criterion has not been met, the harvest target decreases. In either case, the harvest schedule (assignment of harvest activities to strata or stands) is cleared (reset to null, wiped clean), and the scheduling of harvest activities begins again.

Procedurally, a binary search process begins with a harvest target and an *increment*. The increment is an amount that the target will increase when the target is successfully scheduled in all time periods of the time horizon or likewise will decrease when the target is unsuccessfully scheduled in one or more time periods of the time horizon. The increment does not change until either (a) the harvest level target has been unsuccessfully scheduled in one or more of the time periods (until an upper bound has been identified in a maximization problem) or (b) until a lower bound has been identified in a minimization problem. At that point, upper and lower bounds have been located, and for each subsequent simulation the range of potential harvest levels is cut in half to define the new target harvest level. One iteration of this form of simulation might employ logic such as this:

```
If (failed once = 1) then
  increment = 0.5 × increment
Endif

If (harvest target met in all periods) then
  If (increment > stopping criterion) then
    harvest target = harvest target + increment
  Endif
Elseif (harvest target not met in all periods) then
  If (increment > stopping criterion) then
```

```
      harvest target = harvest target - increment
    Endif
    failed once = 1
  Endif
```

As illustrated above, each time the increment is cut in half it is compared to the stopping criterion, which is defined by the planner. It essentially represents the size of the increment below which it seems irrelevant to continue the binary search process. For example, if the target is expressed in tons of wood, it would seem unreasonable to continue the binary search process when the increment is reduced to as small as, for example, the tons of wood one could load onto one typical log truck. This may be around 25 tons of wood for planning problems situated in the southern United States.

4 Contemporary Forest Harvest Scheduling Problem

Based on the information provided in Sect. 4 of Chap. 2, assume that a binary search process would be employed to seek the highest even-flow harvest volume that can be scheduled from the southern Arkansas forest over the 15-year time horizon. Some relaxation of constraints is allowed here. For example, the harvest decisions will be fractional in nature, as in a linear programming search process, and therefore harvest adjacency constraints will not be assessed. Timber stands are scheduled for harvest based on the average age of trees (oldest first). Stands that are assigned a partial final harvest (a final harvest applied to less than the total area in the stand) will have their residual forested areas grown forward in time at a rate consistent with how the entire stand would have been grown forward in the absence of a harvest activity. For binary search purposes, the initial target harvest level is 19,000 tons per year. If this harvest level is possible to schedule during each year of the time horizon, the target harvest level increases by 1000 tons per year. These attempted target harvest levels are considered lower bounds on the problem until it is not possible to obtain the target harvest level in any 1 year. The attempted target harvest level associated with the unsuccessful harvest schedule then becomes the upper bound on the problem, and binary search is employed from this point forward to approach the maximum volume that could be scheduled over time. Once the binary search adjustments consist of about one or two truckloads of wood, the process terminates. From employing binary search to this problem, we find the following:

First attempt

Target volume is 19,000 tons (successfully scheduled)
Lower bound = 19,000 tons
Upper bound = undetermined

Second attempt

Target volume is 20,000 tons (not successfully scheduled)
Lower bound = 19,000 tons
Upper bound = 20,000 tons

Third attempt

Target volume is 19,500 tons (successfully scheduled)
Lower bound = 19,500 tons
Upper bound = 20,000 tons

Fourth attempt

Target volume is 19,750 tons (not successfully scheduled)
Lower bound = 19,500 tons
Upper bound = 19,750 tons

Fifth attempt

Target volume is 19,625 tons (successfully scheduled)
Lower bound = 19,625 tons
Upper bound = 19,750 tons

Sixth attempt

Target volume is 19,687.5 tons (successfully scheduled)
Lower bound = 19,687.5 tons
Upper bound = 19,750 tons

Seventh attempt

Target volume is 19,718.75 tons (successfully scheduled)
Lower bound = 19,718.75 tons
Upper bound = 19,750 tons

The forest harvest schedule that produced the last successful attempt to meet the target harvest volume (19,718.75 tons) in each of the time periods in the time horizon would then be reported. Certainly, the decisions employed could have involved integer assignments for the final harvest of timber stands. This type of modelling process would then facilitate an assessment of potential harvest adjacency constraint violations. However, scheduling whole stands for harvest during a binary search process would require scheduling as closely to the target as possible, given some assumptions on proximity to the target, as each timber stand contains a unique amount of volume. Employing a binary search process with these assumptions may lead to scheduled harvest levels that are significantly above the target. Alternatively, the process for selecting timber stands from the sorted list may be designed to ignore the selection rule (oldest first) when the inclusion of the last necessary timber stand may result in scheduled harvest volumes that far exceed the target. The adjusted process might then allow the selection of another timber stand further down the sorted list that results in scheduled volumes closer to the target.

5 Applications

Binary search methods have been employed to assess potential forest harvest levels (i.e., timber supply issues) across very broad areas (Sutfin, 1985). And, binary search has been employed in forest harvest scheduling efforts where both commodity production and financial measures are used as objectives and where relevant operational constraints are employed (Johnson & Tedder, 1983). Models such as ECHO, SIMAC, and TREES were developed for southern and western United States forests to examine the economic returns and sustainable harvest levels that might be obtained when various forest management assumptions are modeled (Sassaman et al., 1972; Walker, 1975; Tedder et al., 1978; Marty, 1979; Tedder & Schmidt, 1980; Schmidt & Tedder, 1981). Although there was no apparent feedback from outcomes of a simulated harvest schedule that might have been used to adjust a target sustainable flow of wood, an age-based rule for scheduling harvests period by period was proposed even earlier by Howard et al. (1966).

In a southern United States forest management problem, Tedder et al. (1978) described an application of two binary search models to the management of a 23,941 acre (9689 ha) forest in Oklahoma primarily populated with loblolly pine (*Pinus taeda*). In a Canadian forest management problem, Chen et al. (2011) combined a goal programming problem that included timber, wildlife, and recreation outcomes with a binary search approach for changing the target level of the goals. The idea was to adjust the goal targets for a 27,354 acre (11,070 ha) forest incrementally to determine whether a more effective allocation of management activities could be devised while meeting the intent of multiple-use forest management.

In a western United States forest management problem, Montgomery et al. (1986) used binary search to assess the impact of fires on sustainable forest harvest levels. In this approach, a binary search algorithm was applied to forest strata to determine a sustainable harvest level. Then, some areas of land from the strata where activities were scheduled for harvest during time period 1 were removed and assumed to be destroyed by fire. Land areas assumed to be affected by fires were then returned to the youngest forest age class strata. Binary search was then employed again to assess the impact of fires on the sustainable harvest level. In another western United States forest management problem, Bettinger et al. (2005) employed binary search to the scheduling of harvesting activities on industrial and state forest lands. In this case, stands, rather than strata, were scheduled for harvest subject to a minimum harvest age and harvest adjacency and green-up constraints. Further, *seed stands* were selected from a sorted list based on their value, then physically adjacent stands (feasible for harvest, regardless of their position in the sorted list) may have been added to a seed stand to build a harvest block in an effort to simulate harvest areas that resemble a desired distribution of sizes.

6 Summary

Of the various approaches one can employ to develop a forest harvest schedule, binary search is among the easiest to implement and should have a low computational cost for each simulation of sustainable harvest levels (Johnson & Tedder, 1983). In general, a timber production objective guides the process, but the approach is not necessarily restricted to this sort of commodity production perspective. A sorted list of activities to schedule is also required, and again, as long as the activities and the land areas to which they relate can be associated directly with the objective, one can use this approach to assess the production of other resources. One of the most unique aspects of binary search is the manner in which forest management activities are scheduled—sequentially by time period, from the first to the last. The other approaches that are discussed in this book allow the scheduling of harvest activities, for better or for worse, during any time period within the time horizon, with no restriction on sequencing. However, given the sequential scheduling strategy, a relatively simple commodity production objective, a rule employed for selecting forest areas for harvest (e.g., oldest stands first), and procedures for assessing feasibility, the resulting solution (forest harvest schedule) may not be the most efficient (Garcia, 1990). As with other approaches to forest harvest scheduling, the final solution to the problem being solved is an approximation of the optimal solution. In other words, with this method, optimization of an objective is not guaranteed. Yet as a search process, binary search is often considered as easy to understand by practitioners as Monte Carlo simulation and hill-climbing approaches.

Take Home Message
Binary search is a logical, deterministic, sequential search process that (generally, in forestry) simulates the development of a harvest schedule which results in the highest, and most even harvest levels over time. Wood flow targets can be static, or they can vary and represent a desired pattern of harvest over time. Binary search does not optimize an objective function, rather it simulates a forest harvest schedule based on opportunities available in a sorted list of potential harvest activities. However, typical constraints on the scheduling of harvest activities can be incorporated into the search process to emulate the management of a real-world system.

Questions
1. What conditions of the search behavior of binary search suggest that it does not use a holistic view when solving a problem?
2. Aside from age- or value-related sorting criteria, what other ways can stands of trees be ranked prior to scheduling management activities?
3. What characteristics of binary search might appeal to forest planners when they employ it to develop a forest harvest schedule?

4. Could binary search be employed to develop a forest harvest schedule for today's typical contemporary forest management problems? Why or why not?
5. What are some disadvantages of using binary/integer assignments when scheduling stands or strata within a binary search process?
6. If the target harvest volume for a binary search process was represented in tons, or board feet, or cubic meters, or cubic feet, what stopping rule would seem reasonable to apply?
7. Can a binary search process locate the optimal solution to a contemporary forest harvest scheduling problem?

References

Bettinger, P., & Johnson, K. N. (2003). Spatial scheduling of forest management activities using a dynamic deterministic harvest block aggregation process. *Journal of Forest Planning, 9*(1), 25–34.

Bettinger, P., Lennette, M., Johnson, K. N., & Spies, T. A. (2005). A hierarchical spatial framework for forest landscape planning. *Ecological Modelling, 182*(1), 25–48.

Brodie, J. D., & Sessions, J. (1991). The evolution of analytic approaches to spatial harvest scheduling. In M. A. Buford (Ed.), *Proceedings of the 1991 symposium on systems analysis in forest resources*. (General Technical Report SE-74) (pp. 187–191). U.S. Department of Agriculture, Forest Service, Southeastern Forest Experiment Station.

Bureau of the Budget. (1962). *Automatic data processing ... glossary*. Executive Office of the President, Bureau of the Budget.

Chen, Y.-T., Zheng, C., & Chang, C.-T. (2011). 3-level MCGP: an efficient algorithm for MCGP in solving multi-forest management problems. *Scandinavian Journal of Forest Research, 26*(5), 457–465.

Davis, C. H. (1969). The binary search algorithm. *American Documentation, 20*(2), 167.

Garcia, O. (1990). Linear programming and related approaches in forest planning. *New Zealand Journal of Forestry Science, 20*(3), 307–331.

Halpern, M. (1958). Variable-width tables with binary-search facility. *Communications of the ACM, 1*(2), 1–4.

Hann, D. W., & Brodie, J. D. (1980). *Even-aged management: Basic managerial questions and available or potential techniques for answering them (General Technical Report INT-83)*. U.S. Department of Agriculture, Forest Service, Intermountain Forest and Range Experiment Station.

Howard, R. A., Gould, E. M., & O'Regan, W. G. (1966). Simulation for forest planning. *Simulation, 7*(1), 44–47, 50–52.

Johnson, K. N., & Tedder, P. L. (1983). Linear programming vs. binary search in periodic harvest level calculation. *Forest Science, 29*(3), 569–581.

Katz, D. L., Organick, E. I., Navarro, S. O., & Carnahan, B. (1961). *Second annual report of the project on the use of computers in engineering education*. University of Michigan, College of Engineering.

Lesuisse, R. (1983). Some lessons drawn from the history of the binary search algorithm. *The Computer Journal, 26*(2), 154–163.

Lyon, K. S., & Sedjo, R. A. (1986). Binary-search SPOC: An optimal control theory version of ECHO. *Forest Science, 32*(3), 576–584.

Manna, Z., & Waldinger, R. (1987). The origin of binary search. *Science of Computer Programming, 9*(1), 37–83.

Marty, R. J. (1979). *Options for calculating allowable timber harvest.* (Research Report 371). Agricultural Experiment Station, Michigan State University.

Montgomery, C. A., Brodie, J. D., & Cleaves, D. A. (1986). Allowable cut effect and fire-damage appraisal. *Western Journal of Applied Forestry, 1*(4), 100–103.

Overholt, K. J. (1973). Optimal binary search methods. *BIT, 13*(1), 84–91.

Sassaman, R. W., Hold, E., & Bergsvick, K. (1972). *User's manual for a computer program for simulating intensively managed allowable cut.* (General Technical Report PNW-1). U.-S. Department of Agriculture, Forest Service, Pacific Northwest Forest and Range Experiment Station.

Schmidt, J. S., & Tedder, P. L. (1981). A comprehensive examination of economic harvest optimization simulation models. *Forest Science, 27*(3), 523–536.

Seidl, T., & Enderle, J. (2011). Binary search. In B. Vöcking, H. Alt, M. Dietzfelbinger, R. Reischuk, C. Scheideler, H. Vollmer, & D. Wagner (Eds.), *Algorithms unplugged* (pp. 5–11). Springer-Verlag.

Sessions, J., Johnson, D., Ross, J., & Sharer, B. (2000). The Blodgett plan: An active-management approach to developing mature forest habitat. *Journal of Forestry, 98*(12), 29–33.

Sutfin, T. W. (1985). *An evaluation of two timber supply models for state assessments.* (Master of Forestry thesis). Oregon State University.

Tedder, P. L., & Schmidt, J. S. (1980). Allowable cut effect and the ECHO algorithm. *Forest Science, 26*(3), 374–376.

Tedder, P. L., Weaver, G. H., & Kletke, D. D. (1978). Timber RAM and ECHO: Alternative harvest planning models for a southern forest. *Southern Journal of Applied Forestry, 2*(3), 98–103.

Walker, J. L. (1975). ECHO: Solution technique for a nonlinear economic harvest optimization model. In J. Meadows, B. Bare, K. Ware, & C. Row (Eds.), *Systems analysis and forest resource management* (pp. 172–188). Society of American Foresters.

Chapter 6
Hill-Climbing Search

Learning Objectives

The objectives of this chapter on hill-climbing search are to help the reader

- Understand the history and motivation of the hill-climbing search process
- Understand the general manner in which deterministic hill-climbing search operates
- Understand the general manner in which stochastic hill-climbing search operates
- Understand the main limitations associated with the use of hill-climbing search

1 Introduction

An exhaustive search for the solution to a management problem would involve assessing every potential solution for both quality (as reflected in the objective function value) and feasibility (as reflected in the assessment of constraints). *Complete enumeration* of all possible outcomes may, however, be computationally expensive and require an amount of time unacceptable to the person or organization seeking the answer, the optimal solution to a problem. In lieu of an exhaustive search for the optimal solution to a problem, all of the search processes described in this book can be characterized as using *local search* strategies, and some of them can arguably be considered a *greedy search* for a global optimum solution to a problem. A greedy search algorithm selects the best actions possible at each iteration (whether building a solution from the null set or adjusting a previously built solution) without visiting solutions of lower quality as the search progresses (Sammut, 2017). It has been suggested that these types of algorithms have been developed in response to the time-sensitive nature of solving complex problems (Chen et al., 2007), where search processes attempt to examine only a small set of alternative solutions to a problem (Feo & Resende, 1995). The low level of complexity of these algorithms and the resulting search efficiency have been noted as reasons why people employ these algorithms to solve difficult problems (Dor, 1997).

P. Bettinger, *Forest Harvest Scheduling*, Springer Texts in Business and Economics,
https://doi.org/10.1007/978-3-031-89432-9_6

It is acknowledged here that the separate application of greedy search algorithms and hill-climbing search processes may differ by domain application (forestry, transportation, communications, etc.), and thus the concepts may appear to be technologically different when viewed from different domain perspectives (Wilt et al., 2010). However, the treatment of these algorithms in this book, with reference to forest management planning and the development of a strategic or tactical forest plan, is the same. These are search processes that move from one local feasible solution (in the solution space) to another local feasible neighbor without considering prior history of movement through the solution space and without thinking about what to do after a move has been made (Russell & Norvig, 2010). In comparison to other search processes described in this book, hill-climbing search processes consist of general problem-solving procedures (Doran & Michie, 1966) which are sometimes considered stubborn, yet rational approaches to solving problems.

In general, local search is a trial-and-error process that is perhaps the oldest optimization method (Grover, 1992). As was noted above, hill-climbing search fits this characterization only when improvements in the objective function value are allowed. The essence of hill-climbing search is that given a current solution, the only changes that can be made to it are those that lead to a better quality, feasible solution. Perhaps these changes are made randomly or perhaps they are made deterministically (the best changes first). In either event, the search progresses until there are no more improvements to be made. The search process therefore involves a sequence of choices that are made because they improve the quality of a solution (Qiang et al., 2020) whether the choices are made randomly or deterministically. Unfortunately, while logical and perhaps rational, this type of search process may become stranded or stuck at a local optimum solution when there are no more moves to be made in the solution space that can improve the quality of the solution.

2 Background

With no other caveats to the search process described here, iterative improvement search processes such as hill-climbing algorithms move from a current feasible solution to a neighboring feasible solution that is higher in quality with respect to the objective function value (Nurmela, 1993). This is why some have called them greedy search algorithms. Given a current feasible solution, a general hill-climbing algorithm would create a candidate new solution from the neighborhood of potential solutions, and (a) if the objective function value of the candidate new solution is higher in quality and (b) no constraints are violated, the candidate solution replaces (and becomes) the current solution to the problem (Johnson & Jacobson, 2002). This type of iteration of the search process (select a candidate solution from the neighborhood, assess the candidate solution, potentially accept the candidate solution) is repeated until there are no more candidate solutions that will improve the objective function value of the current solution. If the potential move is informed by its potential impact, a similar search process, the steepest descent algorithm, uses knowledge of the impact of potential changes to the objective function value of the current solution, and chooses the move that results in the greatest reduction of impact.

These types of search processes are, in computer programming language parlance, loops that continue to improve the value of a solution until there are no more improvements available (Russell & Norvig, 2010). Therefore, these search processes terminate when no improvements (upward in the case of a maximization problem and downward in the case of a minimization problem) in an objective function value can be made, and the termination point (final feasible solution) is always considered a local optimum solution (Nurmela, 1993) because these forms of local search processes do not guarantee global optimality has been achieved (Johnson & Jacobson, 2002). However, some have suggested that the best solution generated by processes such as these may reasonably approximate an optimal solution under certain conditions (Korsh, 1981).

The basic implementation of a hill-climbing search process therefore involves:

1. Selecting a feasible solution (or null set) from the solution space and labeling it the current solution to the problem.
2. Temporarily transforming the current solution to a new feasible solution in the solution space, temporarily moving from the current solution to a local neighbor.
3. Labeling the temporary solution as the current solution and storing it in memory as the best solution if the local neighbor is feasible with respect to the constraints and has a more favorable objective function value than the current solution from which it came.

3 Process

There are two general uses of hill-climbing algorithms for assisting with the task of forest harvest scheduling: (1) in developing a harvest schedule that begins with a null or empty set, or (2) in adjusting a previously defined full schedule of harvest activities. From a random starting point (randomly developed, feasible harvest schedule), a hill-climbing search algorithm may utilize an assessment of all potential neighboring solutions before one change to the current solution of a problem is selected. This neighboring solution can be reached from the current solution by a 1-opt, 2-opt, or n-opt move. Assessing all potential changes to the current solution can be computationally expensive, as many neighbors need to be assessed (their potential objective function values have been calculated) before one is selected. In this sense, the hill-climbing search process is deterministic in nature, much like tabu search (Chap. 10) without the ability to avoid becoming stuck at a local optimum. Alternatively, GRASP (greedy randomized adaptive search procedures) is a hill-climbing search process that assesses, ranks, and then selects one of the top candidates from a restricted candidate list (Feo & Resende, 1995). With each iteration of the search process the potential benefits of each subsequent random choice need to be assessed before one is selected, as their utility toward the value of the harvest schedule may change based on other previously selected activities.

Stochastic hill-climbing search processes have an advantage over deterministic greedy search processes in that fewer neighbors from the neighborhood might be assessed before one is deemed acceptable. Unlike the deterministic hill-climbing search process, a stochastic hill-climbing process is not necessarily rational, but it may initially be computationally faster. And unlike a deterministic hill-climbing search process, a stochastic hill-climbing process may encounter difficulties in knowing when a search should terminate. In a deterministic hill-climbing search, once all of the neighbors from the local neighborhood have been assessed for fitness (their potential objective function values have been calculated), it is relatively easy to know that the end of the search has been encountered—none of the local neighbors will result in an objective function value that is higher in quality than the current solution's objective function value. With a stochastic hill-climbing search process, one may have to employ a separate tracking process to understand whether all local neighbors have been assessed and whether there are no more moves in the solution space that can be made to improve the quality of the solution.

3.1 General Description

With *s*-metaheuristics in particular, neighboring solutions are the ones that can be reached through 1-opt or *n*-opt changes to a current solution. In this type of local search process, a great portion of the character of the current solution will be found in the temporary, proposed solution. The search process movement around and through the solution space may therefore be rather limited with each iteration of a search process. In developing a plan of action for the management of a forest, a basic process (Fig. 6.1) for a hill-climbing algorithm that seeks to maximize an objective may involve the following:

- Develop an initial feasible integer solution to the problem (Step 1). This becomes both the current and the best solution to the problem, by default. The initial feasible solution can be developed randomly or deterministically or may simply be represented by a null (empty) set of management actions. For research purposes, a random initial solution is generally preferred.
- Propose a deterministic or random change to the current solution, creating a temporary proposed solution (Step 2).
- Determine whether the temporary proposed solution is feasible with respect to all constraints (Step 3).
  - If the temporary proposed solution is feasible, determine whether it will have an objective function value that is higher in quality than the current solution to the problem (Step 4), and if so, make it the current solution (Step 5).
  - Otherwise, discard this attempt to change the current solution.
- Determine whether it is time to terminate the search (Step 6).

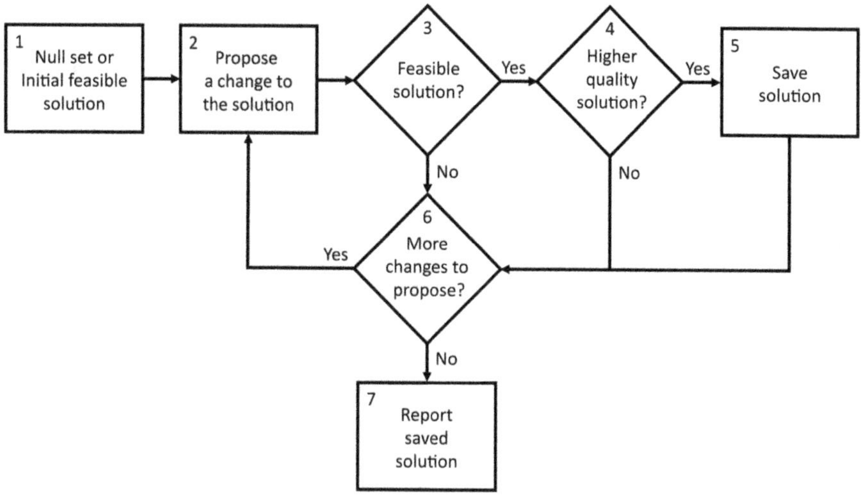

Fig. 6.1 A basic hill-climbing or greedy search process for forest harvest scheduling. Credit: Pete Bettinger

When a hill-climbing search process is employed, Step 5 in Fig. 6.1 would only occur when the temporary, proposed solution has a higher quality objective function value than the current solution. Further, each new solution generated would technically be considered the best solution, and there would be no reason to store the solution in memory. When a hill-climbing search process is employed, there will also come a point in time when no further improvements are possible within the local neighborhood. In effect, the search process will find the best solution in a local neighborhood. What is not known to the hill-climbing search process is that there may have been a better solution somewhere else in the solution space. Reaching it might have been impossible from the starting point of the search process. To illustrate this point, Fig. 6.2 illustrates five independent attempts using the same hill-climbing algorithm to develop a forest harvest schedule that maximizes the net present value of activities subject to final harvest adjacency and green-up restrictions and wood flow restrictions.

3.2 *Selection of Moves*

During each iteration of a hill-climbing model, a potential new solution is selected from the local neighborhood, in mathematical space, of potential solutions surrounding the current solution. The current solution plays a dual role in hill-climbing, both as the basis for a transition to a new solution, and as the criterion for whether the next solution should be accepted or discarded (Newell, 1969). If the potential new solution has an objective function value higher in quality than the current solution,

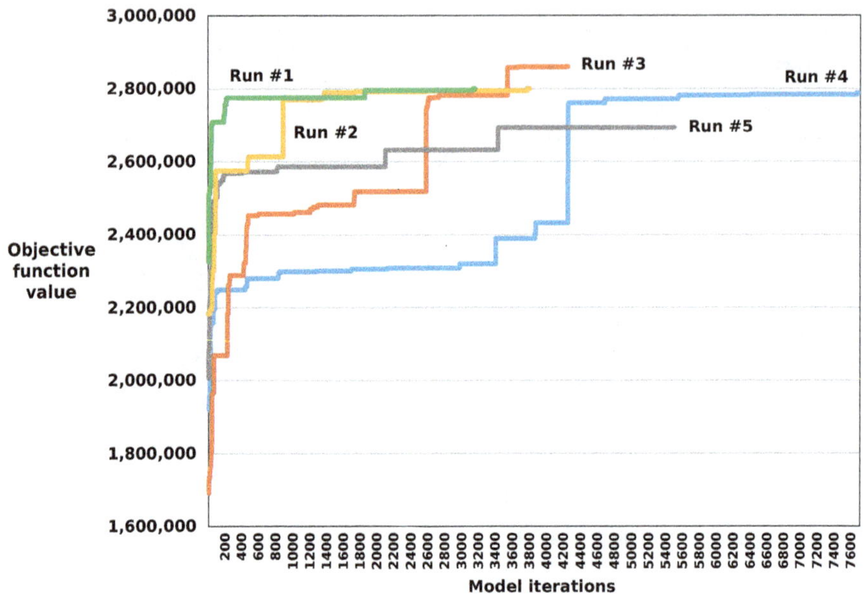

Fig. 6.2 Results from five independent attempts to develop a forest harvest schedule using a stochastic hill-climbing search process. Credit: Pete Bettinger

it becomes the current solution, otherwise it is rejected, and other neighbors are evaluated. This process continues until there are no more neighbors to assess. A basic hill-climbing search process retains no memory of past operations and assesses no broad-scale knowledge of the impact of potential future choices (Russell & Norvig, 2010). The only caveat to the latter part of this statement is that there may be times when an assessment of several local (nearby, in the solution space) potential changes to a feasible solution to a problem might be conducted prior to selecting the change that is made within each iteration of the search process.

With *stochastic hill-climbing*, the probability of selecting potential moves from the local neighborhood in the solution space may be uniform or may vary depending upon the ability of proposed choices to contribute to the objective (which may require analysis). With *deterministic hill-climbing*, a search process would ideally assess all potential moves before selecting one (or terminating the search). If two or more potential moves result in the same improvement in objective function value, perhaps some arbitrary rule would be devised to select one of them (Doran & Michie, 1966).

With respect to both types of hill-climbing search described here, stochastic and deterministic, the constraints are usually assessed (Step 3 in Fig. 6.1) once a local neighbor solution has been selected. For complex problems, it may be computationally expensive to assess the constraints associated with every potential local neighbor (in the deterministic case) prior to selecting one from the local neighborhood.

Example

As an example of a deterministic hill-climbing search process, imagine there are five timber stands (Fig. 6.3), and each contains mature timber that can be harvested and sold. Assume there are three time periods in the time horizon, and that each is 5 years long. Any of the five timber stands can be scheduled for harvest during any of the three time periods, but the assignment is discrete (i.e., a stand can only be harvested during one time period, and completely harvested when the action occurs). The non-discounted value of the timber is illustrated in Table 6.1, along with the present value of the timber using a 5% discount rate (values are discounted in the middle of each time period). For the entire forest, assume the objective function is designed to maximize the net present value. Further, assume that according to organizational policy, two adjacent stands cannot be assigned for harvest within the same time period.

A deterministic hill-climbing search process would select first the choice that best contributes to the objective function. This selection involves scheduling stand 1 for harvest during time period 1. Since this is the best option for the stand, no other options are considered. Of the remaining stands, harvesting stand 4 during time period 1 would best improve the objective function value. Since stand 1 is not adjacent to stand 4, this assignment is feasible. Of the remaining three stands, harvesting stand 3 during time period 1 is the most attractive, but stand 3 is adjacent to both stands 1 and 4, which are already scheduled for harvest during time period 1. Therefore, this assignment is not possible. The opportunity to harvest stand 3 during time period 2 is the next best option and leads to a feasible solution, so this assignment is made. Stands 2 and 5 remain unscheduled at this point. Stand 5 has greater potential to improve the objective function value, but since it is adjacent to stands 3 and 4, stand 5 can only be harvested during time period 3, so this assignment is made. Stand 2 faces a similar situation—it can only be assigned a harvest during time period 3 since it is adjacent to stands 1 and 3. The resulting schedule is therefore:

Time period 1	Harvest stand 1 ($88,517.01) and stand 4 ($57,536.06)
Time period 2	Harvest stand 3 ($43,485.83)
Time period 3	Harvest stand 2 ($11,982.36) and stand 5 ($26,576.38)
Total net present value	$228,097.65

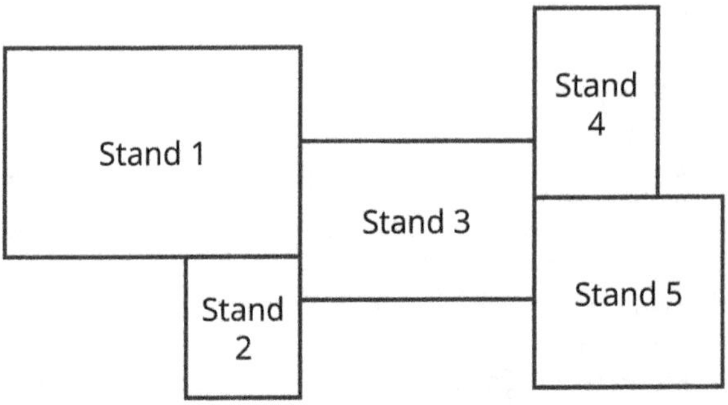

Fig. 6.3 A small, hypothetical forest containing five timber stands of different sizes. Credit: Pete Bettinger

Table 6.1 The non-discounted revenue obtainable from harvesting five timber stands of different sizes from the hypothetical forest, along with the present value equivalents when a 5% discount rate is employed. Credit: Pete Bettinger

		Time period	
Stand	1	2	3
	Non-discounted potential revenue (US$)		
1	100,000.00	104,000.00	108,160.00
2	20,000.00	21,000.00	22,050.00
3	60,000.00	62,700.00	65,208.00
4	65,000.00	67,600.00	70,304.00
5	45,000.00	47,025.00	48,906.00
	Discounted potential revenue ($USD)		
1	88,517.01	72,129.61	58,776.06
2	17,703.40	14,564.63	11,982.36
3	53,110.21	43,485.83	35,435.18
4	57,536.06	46,884.25	38,204.44
5	39,832.66	32,614.38	26,576.38

3.3 *Termination of the Search*

Hill-climbing search processes generally terminate when there are no more alternatives (changes to a solution) that will lead to one that is higher in quality than the previous solution. One of the drawbacks of using hill-climbing algorithms is their potential to terminate the search process too early (Wilt et al., 2010) when the search process encounters mathematical ridges, plateaus, or otherwise local optima within the solution space, and due to the inability of the search process to select inferior quality moves (Russell & Norvig, 2010). There may be instances when a hill-climbing search process is designed to terminate after a maximum assumed number of model iterations have been conducted, or when a solution of a predefined quality has been located (Feo & Resende, 1995). With stochastic hill-climbing search, additional logic may be necessary to understand how many options to improve the solution are available at each iteration of the search process in order to terminate in a timely manner. Otherwise, the search process may simply continue to attempt randomly selected moves in the solution space ignorant of the fact that none of them will improve the quality of the current solution.

3.4 *Random Restart*

NP-hard problems may have numerous local optima on which hill-climbing algorithms might terminate (Russell & Norvig, 2010). Since one is generally unsure whether a local optimum encountered is in fact the global optimum solution to a problem, a restart mechanism has been proposed for individual runs of a model, where the search process pauses (when search stagnation is sensed) and begins anew in a different place within the solution space. Essentially, within a single run of a hill-climbing search process may encompass many different starting points, as opposed to many different runs of a hill-climbing search process utilizing individual, unique starting points. A random restart hill-climbing search process was used to schedule the harvest period in a stand-level management problem that focused on continuous cover forestry practices in Finland (Rämö & Tahvonen, 2017).

4 Contemporary Forest Harvest Scheduling Problem

Loosely based on the example from Chap. 2, imagine that there are no assignments made at the initiation of a deterministic hill-climbing search process. In other words, the initial solution to the forest harvest scheduling problem is the null set, and the objective function value that maximizes discounted net revenue is $0. From the entire set of potential assignments of final harvest activities to the 31 stands in the southern Arkansas forest, the assignment that contributes the most to the objective function value would be selected first. In this case, it is the assignment of a final harvest during time period 1 to timber stand 2, which contributes $382,680 to the objective function value (see Appendix C). Of course, all potential assignments would need to have been evaluated prior to selecting this option. This assignment concludes the first iteration of the deterministic hill-climbing search process.

The second iteration of the deterministic hill-climbing search process could involve changing the assignment of the final harvest to stand 2 to time period 2, because this is the next highest option in value, yet in doing so, the objective function value would decline, as this option is worth $16,893 less than the previously selected option. In fact, any other options for stand 2 are worth less in value than the first option selection, thus one might ignore any other assignments that relate to stand 2. The second iteration of the deterministic hill-climbing search process could then involve the assignment of a final harvest during time period 1 to stand 27, assuming that there are no constraint violations in doing so. Stand 2 and stand 27 are not adjacent neighbors in that they do not share an edge in the physical environment; therefore; these constraints are not violated. This assignment is made, adding $348,225 to the objective function value. Like stand 2, any other options for stand 27 are worth less in value, thus one might ignore other options for stand 27.

The third iteration of the deterministic hill-climbing search process could involve the assignment of a final harvest during time period 1 to stand 1, because this is the next highest option in value ($318,809). However, stand 1 and stand 2 are physically adjacent. Therefore, this assignment, which would allow two adjacent timber stands to undergo a final harvest during the same time period, would be rejected and the next best choice would be made. The logic of the deterministic hill-climbing search process continues in this manner until there are no more possible assignments of harvest activities to timber stands that would increase the objective function value.

In a stochastic hill-climbing search process, the assignment of final harvest activity to a timber stand would require randomly selecting the timber stand, then randomly selecting the potential time period for a final harvest. Assessments of the final harvest adjacency constraints and potential improvement in the objective function value would then be conducted. While prior knowledge of potential improvements in the objective function value is unnecessary, other logic may be required to determine the point during the search whereby no other random selection of timber stand and harvest period could possibly improve the objective function value.

5 Applications

Various works have utilized the capabilities of hill-climbing methods for locating efficient solutions to forestry problems. For example, Chen et al. (2023) described the development of an interesting tree-level optimization process that attempts to adjust (through tree removal) the spatial structure of a dawn redwood (*Metasequoia glyptostroboides*) plantation located in southeast China. Mathematical functions describing tree aggregation and stand openness are maximized, while functions describing tree competition and size relationships among neighboring trees are minimized. Although it is suggested that the approach described is a Monte Carlo simulation method for optimizing the spatial structure of a stand of trees, in an examination of the change in objective function value during the search process, perhaps the process is better described as a stochastic hill-climbing algorithm. In essence, during one iteration of the process, it seems that a temporary random adjustment to the spatial structure of the plantation is proposed, and if the objective function value improves, the temporary adjustment is accepted. The process continues then for a pre-defined number of iterations.

Liu et al. (2006) described a long-term (100-year) forest-level planning problem that involved a 17,381 acre (7034 ha) British Columbia forest. Here, a stochastic hill-climbing algorithm was employed to maximize total wood volume scheduled for harvest, subject to minimum harvest age, harvest adjacency and green-up, and wood flow constraints. The algorithm attempted to use feasible, yet random changes to a harvest schedule to increase the value of a forest plan. Liu et al. (2006) noted that a major disadvantage of this process is that it may ultimately terminate at a local optimum solution in the solution space of the problem with no chance of moving to higher quality areas. They also noted that the performance of the hill-climbing algorithm may be related to the starting point in the solution space, and therefore to improve overall performance, the hill-climbing algorithm should be applied, through independent runs of the model, using a large number of initial, feasible, starting positions within the feasible region of the solution space.

6 Summary

As was suggested earlier in this chapter, a hill-climbing search process will more than likely terminate at a local optimum solution, rather than the global optimum solution to a problem. A general hill-climbing search process may alternatively locate parts of the solution space that are composed primarily of flat regions (tables, mesas) where little to no improvement in solution quality can be obtained, a situation considered the *Mesa Phenomenon* by Minsky (1980). Unfortunately, greedy algorithms such as these may not always perform well, as the starting location of the hill-climbing process within the feasible region of the problem's solution space may affect the ending quality of the final solution (Michalewicz & Fogel, 2002). However, a hill-climbing search process is somewhat rational in nature in that moves throughout a solution space are only deemed acceptable if improvements are made in the objective function value of the problem. On another note, this type of search process might also be characterized as ignorant or naïve because options that may lead to better solutions are ignored simply because one or more decisions along the way do not represent improvements to the quality of the resulting solutions even though they may be necessary. This type of forest harvest scheduling search process might therefore be subject to a *Principle of Standards*.

> *In moving around a solution space, when solution adjustment standards are too high (disallowing a change to a current feasible solution because the objective function value does not improve or because only a small decline in the objective function value is allowed) a search process may become stuck at a local optimum.*

Improvements to hill-climbing search processes include strategies that allow or accept moves that do not necessarily improve the objective function value of a management problem. In future chapters, threshold accepting, simulated annealing, tabu search, and others are in some sense greedy, hill-climbing heuristics, but they include mechanisms that allow the Markov chain of events (which generally assume that future states are conditionally independent of past states [Clarke & Disney, 1985] or more specifically a future state only depends on the current state [Gilks et al., 1996]) to accept moves within the solution space that are composed of lower quality solutions, in an attempt to break free or to avoid becoming lodged in local optima (Chen et al., 2007).

Take Home Message

Hill-climbing search processes are logical search processes that lead somewhat quickly to a local optimum within the feasible region of the solution space. These search processes can employ deterministic selection of changes to a solution to improve its overall quality. However, only changes that improve the quality of the solution are acceptable, and when no more changes are available to improve the quality of the solution, the hill-climbing search terminates. These search processes can alternatively employ stochastic or random selection of changes to a solution to improve its overall quality. Again, only changes that improve the quality of the solution are acceptable, yet an understanding of how many more changes are available to improve the quality of the solution at each iteration of the search process may require additional logic so that this type of hill-climbing search will terminate.

Questions

1. How might a hill-climbing search process be initiated? Would this bias the scheduling effort?
2. If a hill-climbing search process were applied to a problem with an objective function to be minimized, would a different name for it be appropriate, or does *hill-climbing* suffice?
3. If a hill-climbing search process employs a random restart procedure, where the search process pauses and begins anew in a different place within the solution space, in your opinion, how much of the current solution prior to the restart should be altered?
4. With respect to the search process termination rule, how might a stochastic hill-climbing search process and a deterministic hill-climbing search process differ in this regard?

References

Chen, S., Miura, K., & Razzaqi, S. (2007). Analyzing the role of "smart" start points in coarse search-greedy search. In M. Randall, H. A. Abbass, & J. Wiles (Eds.), *Progress in artificial life*. (Lecture Notes in Computer Science, 4828) (pp. 13–24). Springer-Verlag.

Chen, C., Zhou, L., Li, X., Zhao, Y., Yu, J., Lv, L., & Du, H. (2023). Optimizing the spatial structure of Metasequoia plantation forest based on UAV-LiDAR and backpack-LiDAR. *Remote Sensing, 15*(16), Article 4090.

Clarke, A. B., & Disney, R. L. (1985). *Probability and random processes: A first course with applications* (2nd ed.). John Wiley & Sons, Inc.

Dor, A. (1997). The greedy search algorithm on binary vectors. *Journal of Algorithms, 27*(1), 42–60.

Doran, J. E., & Michie, D. (1966). Experiments with the graph traverser program. *Proceedings of the Royal Society A, 294*(1437), 235–259.

Feo, T. A., & Resende, M. G. C. (1995). Greedy randomized adaptive search procedures. *Journal of Global Optimization, 6*(2), 109–133.

Gilks, W. R., Richardson, S., & Spiegelhalter, D. J. (1996). *Markov chain Monte Carlo in practice*. Chapman & Hall.

Grover, L. K. (1992). Local search and the local structure of NP-complete problems. *Operations Research Letters, 12*(4), 235–243.

Johnson, A. W., & Jacobson, S. H. (2002). On the convergence of generalized hill climbing algorithms. *Discrete Applied Mathematics, 119*(1–2), 37–57.

Korsh, J. F. (1981). Greedy binary search trees are nearly optimal. *Information Processing Letters, 13*(1), 16–19.

Liu, G., Han, S., Zhao, X., Nelson, J. D., Wang, H., & Wang, W. (2006). Optimisation algorithms for spatially constrained forest planning. *Ecological Modelling, 194*(4), 421–428.

Michalewicz, Z., & Fogel, D. B. (2002). *How to solve it: Modern heuristics*. Springer-Verlag.

Minsky, M. (1980). Appendix B steps toward artificial intelligence. In *Machine intelligence and robotics: Report of the NASA study group, Final report*. National Aeronautics and Space Administration, Study Group on Machine Intelligence and Robotics.

Newell, A. (1969). Heuristic programming: Ill-structured problems. In J. S. Aronofsky (Ed.), *Progress in operations research, volume III, relationship between operations research and the computer* (pp. 361–414). John Wiley & Sons, Inc.

Nurmela, K. J. (1993). *Constructing combinatorial designs by local search*. Helsinki University of Technology, Digital Systems Laboratory. Series A: Research Reports. No. 27.

Qiang, X., Appiah, M. Y., Boateng, K., & Appiah, F. V. (2020). Route optimization cold chain logistic distribution using greedy search method. *Opsearch, 57*(4), 1115–1130.

Rämö, J., & Tahvonen, O. (2017). Optimizing the harvest timing in continuous cover forestry. *Environmental and Resource Economics, 67*(4), 853–868.

Russell, S. J., & Norvig, P. (2010). *Artificial intelligence, a modern approach* (3rd ed.). Prentice-Hall.

Sammut, C. (2017). Greedy search. In C. Sammut & G. I. Webb (Eds.), *Encyclopedia of machine learning and data mining* (2nd ed., p. 596). Springer.

Wilt, C., Thayer, J., & Ruml, W. (2010). A comparison of greedy search algorithms. In *Proceedings of the Third Annual Symposium on Combinatorial Search (SOCS-10)* (pp. 129–136). Association for the Advancement of Artificial Intelligence.

Chapter 7
Random Search

Learning Objectives

The objectives of this chapter on random search are to help the reader

- Understand the history and motivation of random search processes
- Understand the general manner in which a random search operates
- Understand the use and value of search knowledge and how it might help a random search process
- Understand the main limitations associated with the use of random search

1 Introduction

For millennia, humans have likely been using random search methods to solve problems of importance. Efforts to formalize these procedures began within the last 100 years or so. *The problem of the random walk*, referencing the probability of ending up at a certain point after a certain amount of time by following a fixed distance across a random direction, was offered as a question by Pearson (1905) in a very short, 7-sentence piece in *Nature*, much like questions are now posted in ResearchGate. Monte Carlo simulation, a form of statistical (or stochastic) sampling, was perhaps first employed by Italian–American scientist Enrico Fermi (Metropolis, 1987; Benov, 2016) in scientific efforts he conducted in the 1930s. However, Fermi's colleagues Stanislaw Ulam (a Polish–American scientist) and John von Neumann (a Hungarian–American scientist) were credited as *inventors* of the Monte Carlo method in 1946 because they formalized the process through publication (Gubernatis, 2005). The term *Monte Carlo simulation* was coined by Nicholas Metropolis (a Greek–American scientist) in reference to an uncle of Ulam, who would borrow money from members of his family and use it to gamble in Monte Carlo (Metropolis, 1987). Monte Carlo, of course, is a tourist resort, gambling center, and commune of the constitutional monarchy of Monaco, located along the Mediterranean Sea near the border of France and Italy (Merriam-Webster, Inc.,

P. Bettinger, *Forest Harvest Scheduling*, Springer Texts in Business and Economics,
https://doi.org/10.1007/978-3-031-89432-9_7

2001). With respect to operations research, a Monte Carlo simulation could be considered any method that obtains by random sampling an estimate of a desired outcome, by randomly selecting feasible activities in efforts to model real-world systems (Householder, 1959). In comparison to other purposeful and biased search processes, random search or Monte Carlo simulation stumbles through the solution space hoping to locate a high quality solution. As a search process it may be useful when the creation of an exact solution is impossible due to non-linearities of a problem (McCracken, 1955; Bauer, 1958), or when solutions to complex problems are needed quickly and with a relatively low level of effort (Iuorno, 1963).

2 Background

Monte Carlo simulation, or random search, relies on a series of repeated random sampling efforts to locate a solution to a problem. This type of search, as the name suggests, relies on a *game of chance* with the expectation that the solution (or final answer) will approximate the global optimum if the sample size is large enough (Bauer, 1958; Iuorno, 1963). In short, Monte Carlo simulation associates random numbers with potential decisions described by numerical processes (Hammersley & Morton, 1954) hoping to locate good solutions to problems. For management activity assignment problems in forestry, random search can take at least four forms to either construct or modify (or both) feasible solutions to harvest scheduling problems. In the first form of random search for solution construction purposes, a search process might rely on the development of trial configurations (solutions) that are created randomly and independently from any others that were developed previously (Černý, 1985), where knowledge of relatively good or bad combinations of management activities (in a forestry sense) is lacking with the development of each new solution (Fig. 7.1). For forest harvest scheduling purposes, this has been characterized as *simple random search* (Bullard & Klemperer, 1984). Assume for the moment that a forest contains 1000 timber stands and there are six potential options to manage each stand. Now imagine that there are 1000 dice, each with six sides of course. One die is related exactly to one timber stand. This form of random search might involve rolling all 1000 dice simultaneously, and the assignment of management activity to each timber stand is based on the value facing upward on each die that relates to each stand. Of course, scheduling management activities in this manner may result in many infeasible solutions if complex forest-level constraints are assessed after the assignments have been made.

A second form of random search (Fig. 7.2) for solution construction purposes might involve rolling each die one at a time, assessing forest-level constraints after each roll, then re-rolling a die when constraints are violated. This type of search can become very inefficient when the same stand and the same assignment are attempted multiple times in the development of a single solution (forest harvest schedule). To enhance this type of random search process, perhaps some knowledge of the stands

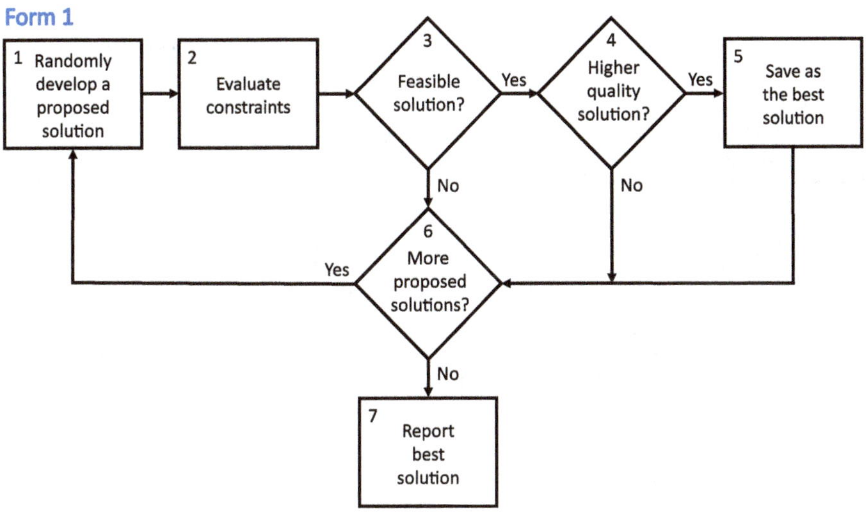

Fig. 7.1 Form 1: A general process for using a simple random search process to construct a solution. Credit: Pete Bettinger

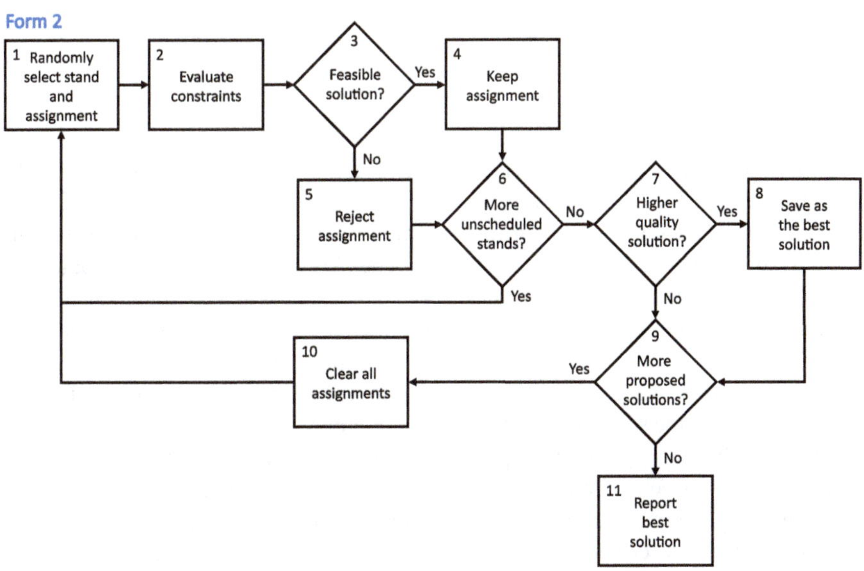

Fig. 7.2 Form 2: A general process for using a random search process that iteratively constructs a solution. Credit: Pete Bettinger

selected is developed, maintained, and consulted to avoid attempting these again during the development of a single solution. For forest harvest scheduling purposes, this may be characterized as *multistage random search* (Bullard & Klemperer, 1984).

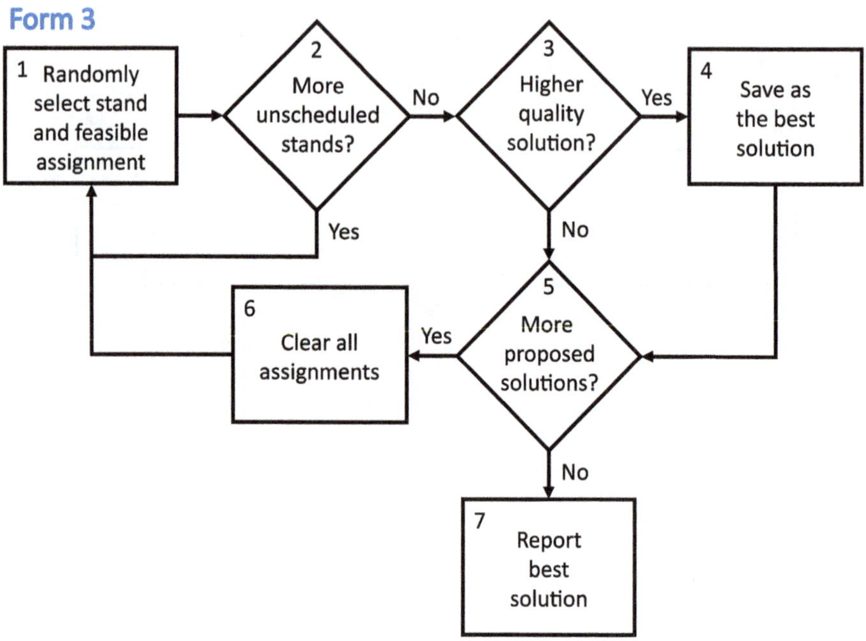

Fig. 7.3 Form 3: A general process for using an iterative random search process to construct a solution only from feasible options. Credit: Pete Bettinger

A third form of random search (Fig. 7.3) for solution construction purposes would be to use knowledge of potential constraint violations to adjust the number of sides each proverbial die contains, ensuring that only feasible harvest schedules will result as each die is rolled. Of course, to enable this type of process, knowledge of potential feasible assignments for each stand needs to be developed, maintained, and consulted to ensure that only feasible assignments are available when a stand is randomly selected. The logic to facilitate this type of enhancement can be very cumbersome to develop; however, once employed, it may allow a more efficient random search of the solution space even though it may correspondingly increase the computation burden by requiring potential feasibility to be assessed continuously. The trade-off between wasting time (developing infeasible solutions) and requiring additional inspection (assessing the feasible options for each stand) is an interesting analysis yet to be described in the literature. As with the previous example, this may also be characterized as multistage random search in forest harvest scheduling (Bullard & Klemperer, 1984).

A fourth form of random search (Fig. 7.4) for solution modification purposes would adjust the assignments of management activities to timber stands for some pre-determined amount of time (or number of iterations). Each adjustment would be random in nature, and each stand therefore might have its assignment randomly changed several times throughout the search process. For example, in Step 1, an initial solution to a forest harvest scheduling problem is created. Perhaps it consists

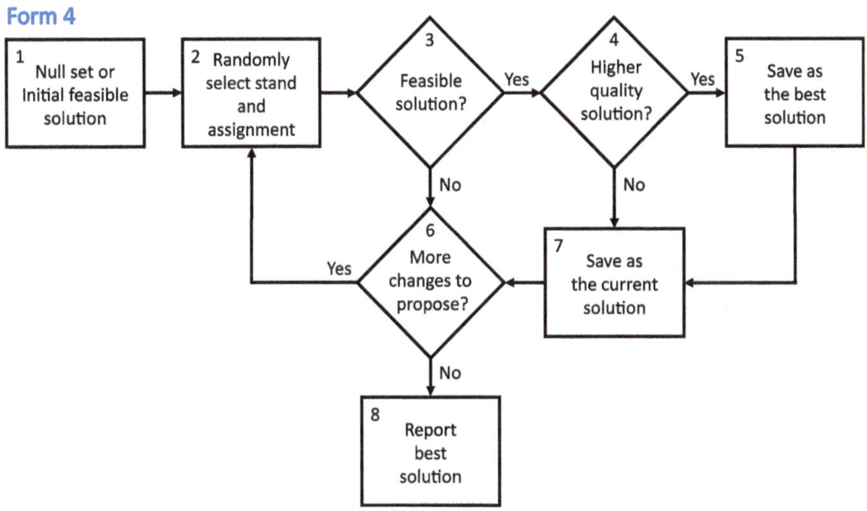

Fig. 7.4 Form 4: A general process for using a random search process to modify a solution. Credit: Pete Bettinger

of a null set of assignments to each timber stand, or perhaps it consists of a random assignment based on one of the first three forms mentioned above. Then (Step 2) a timber stand and a potential management activity assignment are selected randomly, and temporarily inserted into the current solution, creating a proposed solution to the problem. If the proposed solution is not feasible (Step 3), and if there are no more proposed random changes to assess (Step 6), the search process terminates, and the best solution located during the search is reported. If, however, there are more potential changes planned for modifying the current solution, the process continues. If the proposed solution is feasible (Step 3) and results in a higher quality solution (Step 4) with respect to the objective function value, the proposed solution is saved in computer memory as the best solution (Step 5), and it also becomes the current solution to the problem. If the proposed solution does not have an objective function value that is higher in value than the best solution, it simply becomes the current solution (Step 7). Of course, this process could be enhanced in the manner described in Form 3 to increase the efficiency of the search process. Like the previous two examples, this may also be characterized as multistage random search in forest harvest scheduling (Bullard & Klemperer, 1984).

While they may be far from optimal, feasible, randomly generated solutions can generally be developed quickly with a random search process (Brodie & Sessions, 1991). A general process such as this might randomly construct or modify solutions to a problem (Nurmela, 1993), and as a result the process can be considered to have sampled the solution space in a relatively unbiased fashion. However, a general random search process may suffer due to a proposed *Principle of Bias* for forest harvest scheduling efforts.

In a forest harvest scheduling process, bias in the selection of potential moves, or bias in the acceptance of potential moves, is necessary to produce higher quality solutions to complex problems.

As was noted earlier, random search might involve the complete random construction and evaluation of a solution, or the use of random walks through the solution space, moving randomly from one solution to a neighboring solution, in an attempt to improve a single solution (Nurmela, 1993). Through the use of random search, one would hope that navigation through the solution space would be rather free, that it would avoid becoming trapped in local optima, and ideally that it would converge upon the global optimum solution to a management problem (Michalewicz & Fogel, 2002). Random variation during a search process is therefore one diversification mechanism that might be employed to explore new solutions across a broad solution space (Michalewicz & Fogel, 2002). However, rather than accepting only feasible modifications that lead to higher quality solutions, these search processes are different than stochastic hill-climbing search processes in that any feasible modification to the current solution of a problem is acceptable regardless of the impact on the objective function value.

3 Process

Straightforward random searches are uninformed by search history. The hope is that the search process will stumble upon the optimal solution to a problem. Even more to the point, the hope is that the end result of the development of a randomly generated forest plan is feasible with respect to all of the constraints. In conducting this type of search, a computer needs a source of random or pseudo-random numbers from a given uniform or non-uniform distribution (Eckhardt, 1987; Metropolis, 1987; Benov, 2016). And as one might imagine, the more complex a planning problem becomes, the less often a feasible solution might be randomly generated. In the example provided in the previous section, it might be obvious that one run of a random search heuristic could involve scheduling activities to all decision variables simultaneously to construct a solution, or it might involve scheduling activities one at a time in an attempt to improve the quality of an evolving solution to either construct a solution or to modify an existing solution. It was also noted that enhanced random search can be pre-biased or informed through the selection of some management activities that only include feasible assignments (choices) or that contribute most to the objective function value (Brodie & Sessions, 1991), by using a system that weights the selection probability of management activities based on their potential value, or by avoiding the selection of management activities which are known to lead to infeasible solutions.

Example

Random search can theoretically be informed by knowledge developed during the search process. Imagine, for example, that a harvest schedule is being developed one random harvest at a time. This might involve a random selection from a non-uniform distribution the timber stand, the final harvest period, and the intermediate activities that better represent preferred management prescriptions. Further, rather than randomly scheduling activities from the entire population of stands with each iteration of the scheduling model, a random search process could be constrained to schedule activities only from a set of unselected or rarely selected stands. And, once a stand has been selected, the management prescription or harvest period might be randomly selected from a subset of feasible options, rather than all of the options available to a stand. Certainly, these enhancements involve extra work in understanding the available stands and the feasible options for these stands, but one could argue that with these enhancements an informed random search can be enabled. Technically, the scheduling of activities is random in these cases, but for purists, these types of enhanced random search may seem like cheating. Perhaps the most important lesson learned here is that in contrast to uninformed random search, time is not wasted with informed random search.

3.1 General Description

The following discussion focuses on straightforward random search. A distinct search process is emphasized: the construction and modification of an independent, feasible solution to a problem (Form 4 in Fig. 7.4).

1. Develop an initial, feasible integer solution to the problem (Step 1). This becomes both the current and the best solution to the problem, by default. The initial feasible solution can be developed randomly or may simply be represented by a null (empty) set of management actions. As with hill-climbing algorithms, for research purposes a random initial solution is generally preferred.
2. Propose a random change to the current solution (Step 2), creating a temporary proposed solution.
   - Determine whether the temporary proposed solution is feasible with respect to all constraints (Step 3).
     - If the temporary proposed solution is infeasible, determine whether it is time to terminate the search (Step 6).
     - If the temporary proposed solution is feasible, determine whether it will have an objective function value that is higher in quality than the best solution to the problem (Step 4).

If the temporary proposed solution has an objective function value that is higher in quality than the best solution, make it the best solution (Step 5).
Replace the current solution with the temporary proposed solution (Step 7).
Determine whether it is time to terminate the search (Step 6).

In contrast to a hill-climbing search process, each new solution generated would not necessarily be considered the best solution to the problem. There will be points in time when improvements in the objective function value are not possible within the local neighborhood, yet the adjustments are accepted, and the search continues with inferior solutions acting as the current solutions. As with other heuristic methods, this search process cannot guarantee that the optimal solution to a problem will be located. Consistent with a hill-climbing algorithm, what is not known to the search process is that there may have been a better solution somewhere else in the solution space and reaching it might have been impossible from each feasible solution visited.

Example
For a problem that attempted to minimize squared deviations from scheduled harvest volumes and a harvest volume target (a strategy to develop a plan which had nearly even wood flow during each time period of a time horizon), a random search process (Form 2) that simply constructed independent feasible solutions resulted in a set of solutions with a wide variation in quality from one to the next (Fig. 7.5). Similarly, for a problem that attempted to maximize net present value from scheduled harvest activities, a random search process that simply constructed independent feasible solutions again resulted in a set of solutions with a wide variation in quality from one to the next (Fig. 7.6). Using the random construction premise (without random modification), it is obvious from these two examples that the best solution located during the search improved with increasing numbers of independent runs of the model. In construction mode only, a large number of independent runs of a random search model would seem necessary.

3.2 Selection of Moves

In general, whether constructing or modifying a solution using random search processes, the selections of moves (changes in the character of the current solution by adjusting assignments) are purely random in nature. In an uninformed random search, this random assignment of activities to decision variables often leads to the development of infeasible solutions with respect to broader forest-level constraints (e.g., wood flow, harvest adjacency, etc.). An informed random assignment of

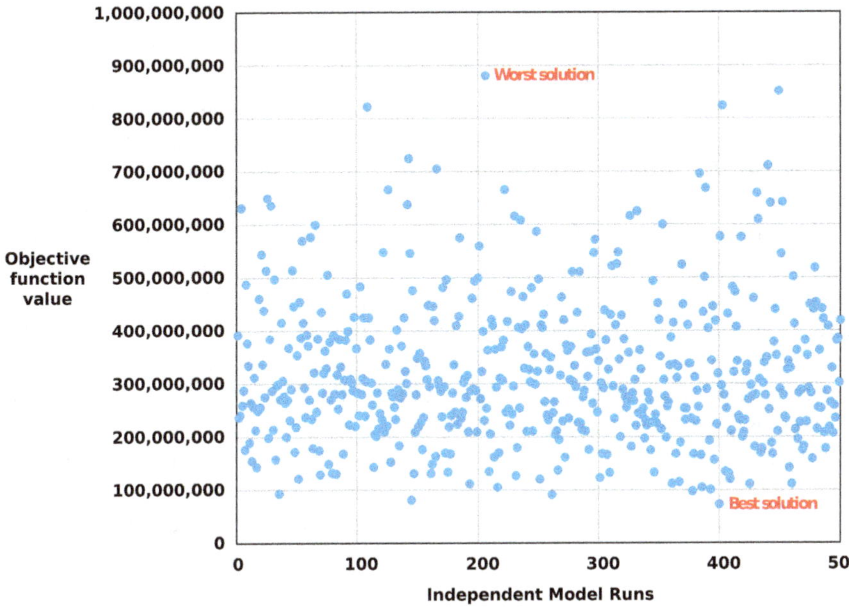

Fig. 7.5 Caption: Objective function values from multiple independent runs, developed solely in construction mode, of a random search model that minimized a wood flow objective. Credit: Pete Bettinger

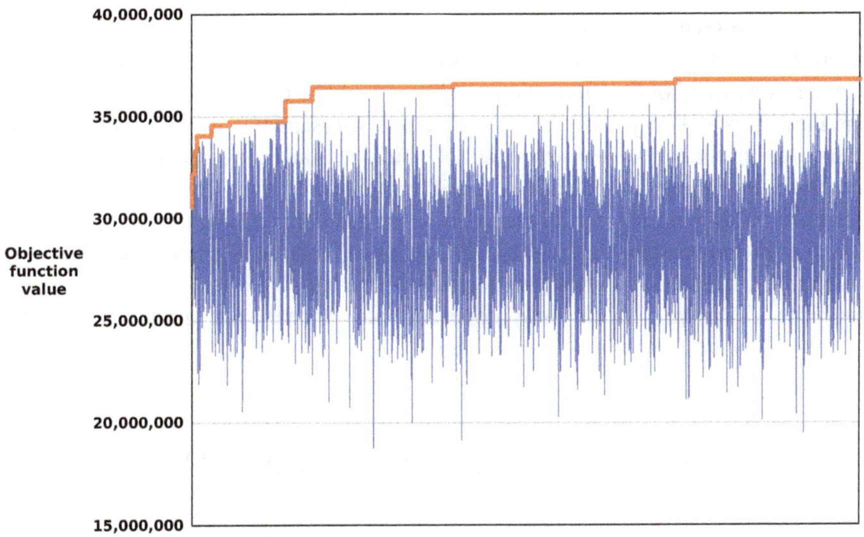

Fig. 7.6 Objective function values from multiple independent runs, developed solely in construction mode, of a random search model that maximized net present value of a forest harvest schedule. Credit: Pete Bettinger

activities to decision variables, as suggested in two of the *forms* noted earlier, can limit the development of infeasible solutions through the moves selected.

After recognizing constraint violations from a potential change to the current solution, actions can be undertaken to repair the violations. On a positive note, at least one variation of this strategy has been shown to potentially be of value (Bettinger & Zhu, 2006). On a negative note, there may be several tactics one can employ to mitigate constraint violations, and these may involve considerable computation effort. For example, if a harvest adjacency constraint is violated by changing the final harvest timing assigned to a timber stand, one could either continue changing the harvest timing of that timber stand (randomly or deterministically) until there are no more constraint violations, or one could change the harvest timing of the adjacent timber stands until there are no more constraint violations. However, if the constraint violated is more complex, such as a wildlife habitat quality constraint that involves assessing the aggregate quality of all patches around the timber stand scheduled for harvest (lowering habitat quality below some minimum desired amount), mitigating the constraint through an adjustment to the harvest assignment of one or more timber stands may be very complicated, if not impossible.

3.3 Termination of the Search

In the case of constructing a solution to a problem, a random search process terminates either when an assignment has been made to all of the decision variables, or when assignments to remaining decision variables are not possible given assignments that have been made to other decision variables. This latter case would suggest that an assessment of *assignment possibilities* would be employed toward the end of the random search to prevent continued attempts to schedule harvest activities (for example) to timber stands when all of the potential assignments will lead to constraint violations. In the case of modifying the solution to a problem, a random search process generally terminates after some pre-defined number of adjustments (successful moves, unsuccessful moves, or both) have been attempted. However, one could also base the termination decision on the length of time (again represented by successful moves, unsuccessful moves, or both) since the last improvement in the quality of the best solution to the problem.

4 Contemporary Forest Harvest Scheduling Problem

As a loose association to the problem presented in Sect. 4 of Chap. 2, in constructing a forest harvest schedule using a random search process (Form 2), one could randomly select a timber stand and randomly select the time period for the final harvest. Constraints would then be assessed. If there are no constraint violations, the assignment of the final harvest time period to the timber stand would be made regardless of the effect on the objective function value. For this example, there is no constraint on the variability of the scheduled harvest volumes. To begin, stand 9 and time period 12 might be selected randomly even though the objective function value ($65,097) is lower than the best opportunity involving this stand (time period 8, $68,256). Memory of this assignment would be noted through the logic of the search process, and no other assignments would be considered for this timber stand. Once each of the remaining timber stands have been assigned final harvest activities (Table 7.1), or assignments of final harvest activities to timber stands are not possible

Table 7.1 A forest harvest schedule constructed through random assignment of final harvest time periods to timber stands. Credit: Pete Bettinger

Randomly selected timber stand	Randomly selected final harvest time period	Contribution to objective function value
9	12	$65,097.06
24	6	180,072.47
11	10	139,144.42
19	10	81,343.44
1	7	275,332.00
26	12	27,112.20
27	1	348,225.12
18	13	75,500.51
6	13	179,587.17
20	12	159,198.36
31	13	52,131.57
13	14	62,235.13
17	7	138,124.96
8	4	273,977.60
10	5	78,820.22
7	1	88,668.61
5	2	111,240.43
12	15	56,461.52
16	2	250,016.62
4	14	152,903.46
25	4	191,773.21
23	0	0.0
14	9	59,914.47
2	11	251,923.13
3	10	177,914.55
21	14	177,165.30
29	12	95,924.15
15	0	0.0
30	12	36,204.00
22	9	144,467.52
28	13	92,784.35
Total		4,023,263.55

due to potential constraint violations (adjacency of harvests, minimum harvest age), the search process terminates. Of course, the solution to a forest harvest schedule developed in this manner could result in volumes scheduled for harvest that are quite uneven since the evenness of scheduled volumes was not controlled.

5 Applications

The quality of forest plans developed using random search processes has been used as a point of comparison with other heuristic methods (Boston & Bettinger, 1999; Bettinger et al., 2002), and often it is concluded that random search methods produce some of the lowest quality solutions to harvest scheduling problems. However, it has been suggested that a random search process may be a valid option to employ in modeling management alternatives for a forest landowner when the objectives of the landowner are uncertain (Johnson et al., 2007).

A process dubbed *Monte Carlo Tree Search* was described by Neto et al. (2020) as a way to develop a forest harvest schedule, and it followed a course of action similar to Form 2 noted earlier in this chapter. The search process was applied to nine different test forests, where the number of stands ranged from 32 to 1363. Two objectives were employed: maximize net present value and maximize a habitat connectivity index. Constraints were imposed on the resulting wood flow, the average ending age of trees in stands, the final harvest sizes, the green-up of final harvest areas, the minimum amount of wildlife habitat area developed or maintained, and the minimum amount of core forest area developed or maintained. The random search process employed allowed Neto et al. (2020) to conclude that there were some instances where solutions (forest plans) contained high levels of habitat connectivity yet had corresponding low net present values.

6 Summary

Given the informed and biased manner in which other algorithms conduct a search, random search may not be the most effective way to locate a high quality solution to a forest harvest scheduling problem. The inefficiency associated with random search is rooted in the fact that while we may want it to explore a certain region of the solution space where high quality solutions may be located, we cannot force it to do so (King, 1951). The manner in which the search is conducted (haphazard adjustments) generally prevents random search from intensifying a search within a high quality area of the solution space, and thus prevents it from locating solutions of greatest value (Michalewicz & Fogel, 2002). On a positive note, random search is a rather easy process to construct, and in many cases, it operates as a relatively fast heuristic for generating feasible solutions to complex problems.

Take Home Message

As its name implies, random search might be considered an irrational search process, yet it provides a measure of diversification that deterministic hill-climbing search processes fail to employ as they move through the solution space while searching for the optimal solution. Random search can be used to construct a feasible forest harvest schedule or used to modify the development of a forest harvest schedule. Unlike hill-climbing search processes, random search will not become stranded at local optima. However, a forest harvest schedule developed using random search may not represent the optimal solution to a problem under investigation. While harvest schedules developed using a random search process may be of low to moderate quality, this information can be of value when assessing the quality of solutions produced by other methods.

Questions

1. Conceptually, everyone has a general feeling about how decisions are made and the processes that are employed. How do you feel about using a random search process to develop a forest harvest schedule?
2. What rule might you devise to terminate a Form 4 (Fig. 7.4) random search process?
3. How might potential changes (moves) made during a random search process be given probabilities of selection that are not uniform?
4. Of the models represented in Figs. 7.1, 7.2, 7.3, and 7.4, which is the most attractive to you for forest harvest scheduling purposes? Why?

References

Bauer, W. F. (1958). The Monte Carlo method. *Journal of the Society for Industrial and Applied Mathematics, 6*(4), 438–451.

Benov, D. M. (2016). The Manhattan project, the first electronic computer and the Monte Carlo method. *Monte Carlo Methods and Applications, 22*(1), 73–79.

Bettinger, P., Graetz, D., Boston, K., Sessions, J., & Chung, W. (2002). Eight heuristic planning techniques applied to three increasingly difficult wildlife planning problems. *Silva Fennica, 36*(2), 561–584.

Bettinger, P., & Zhu, J. (2006). A new heuristic for solving spatially constrained forest planning problems based on mitigation of infeasibilities radiating outward from a forced choice. *Silva Fennica, 40*(2), 315–333.

Boston, K., & Bettinger, P. (1999). An analysis of Monte Carlo integer programming, simulated annealing, and tabu search heuristics for solving spatial harvest scheduling problems. *Forest Science, 45*(2), 292–301.

Brodie, J. D., & Sessions, J. (1991). The evolution of analytic approaches to spatial harvest scheduling. In M. A. Buford (Comp.), *Proceedings of the 1991 Symposium on Systems Analysis in Forest Resources*. (General Technical Report SE-74). pp. 187–191. U.S. Department of Agriculture, Forest Service, Southeastern Forest Experiment Station.

Bullard, S. H., & Klemperer, W. D. (1984). A case for heuristic optimization methods in forestry. *Resource Management and Optimization, 3*(2), 139–147.

Černý, V. (1985). Thermodynamical approach to the traveling salesman problem: An efficient simulation algorithm. *Journal of Optimization Theory and Applications, 45*(1), 41–51.

Eckhardt, R. (1987). Stan Ulam, John von Neumann, and the Monte Carlo method. *Los Alamos Science, 15*, 131–137.

Gubernatis, J. E. (2005). Marshall Rosenbluth and the Metropolis algorithm. *Physics of Plasmas, 12*, Article 057303.

Hammersley, J. M., & Morton, K. W. (1954). Poor man's Monte Carlo. *Journal of the Royal Statistical Society—Series B (Methodological), 16*(1), 23–38.

Householder, A. S. (1959). *A glossary for numerical analysis*. (ORNL-2704). Oak Ridge National Laboratory.

Iuorno, R. F. (1963). *Simulation using the Monte Carlo method*. (Technical Memorandum No. RADC-RAW-TM-63-5). Information Processing Laboratory, Rome Air Development Center, Research and Technology Division, Air Force Systems Command, Griffiss Air Force Base.

Johnson, K. N., Bettinger, P., Kline, J. D., Spies, T. A., Lennette, M., Lettman, G., Garber-Yonts, B., & Larsen, T. (2007). Simulating forest structure, timber production, and socioeconomic effects in a multi-owner province. *Ecological Applications, 17*(1), 34–47.

King, G. W. (1951). Stochastic methods in statistical mechanics. In A. S. Householder (Ed.), *Monte Carlo method*. (Applied Mathematics Series 12) (pp. 12–18). U.S. Department of Commerce, National Bureau of Standards.

McCracken, D. D. (1955). The Monte Carlo method. *Scientific American, 192*(5), 90–94, 96.

Merriam-Webster, Inc. (2001). *Merriam-Webster's geographical dictionary* (3rd ed.). Merriam-Webster, Inc.

Metropolis, N. (1987). The beginning of the Monte Carlo method. *Los Alamos Science, 15*, 125–130.

Michalewicz, Z., & Fogel, D. B. (2002). *How to solve it: Modern heuristics*. Springer-Verlag.

Neto, T., Constantino, M., Martins, I., & Pedrosa, J. P. (2020). A multi-objective Monte Carlo tree search for forest harvest scheduling. *European Journal of Operational Research, 282*, 1115–1126.

Nurmela, K. J. (1993). *Constructing combinatorial designs by local search*. (Series A: Research Reports. No. 27). Helsinki University of Technology, Digital Systems Laboratory.

Pearson, K. (1905). The problem of the random walk. *Nature, 72*(1865), 294.

Chapter 8
Threshold Accepting Search

Learning Objectives

The objectives of this chapter on threshold accepting search processes are to help the reader

- Understand the history and motivation of the threshold accepting search process
- Understand the general manner in which threshold accepting operates
- Understand the use and value of limited diversification in local search options
- Understand the main limitations associated with the use of threshold accepting

1 Introduction

Like other heuristic search processes that follow in this book, a threshold accepting search process utilizes a set of rules that are one step above the move selection and evaluation procedures of hill-climbing and random search processes (Albus, 1981). These heuristic search processes, when encountering a change or potential move in the local neighborhood of the solution space, include procedures for accepting or declining the moves based on recent history of the search process or based on the potential value of the new solution. Threshold accepting (or threshold search) is a stochastic local search process that was initially presented in the literature by Dueck and Scheuer (1990). As a search process for addressing complex problems, threshold accepting has been described as:

- A point-based or *s*-metaheuristic (Talbi, 2009)
- A unidirectional approximation algorithm (Siedentopf, 1995)
- A refined local search algorithm (Winkler & Fang, 1997)
- A convergent algorithm (Lin et al., 1995)

A threshold accepting search process maintains only two feasible solutions to a problem in the memory of a computer during the search. One solution is considered the *current solution*, and the other solution is the *best solution*. Through one of many

P. Bettinger, *Forest Harvest Scheduling*, Springer Texts in Business and Economics,
https://doi.org/10.1007/978-3-031-89432-9_8

(thousands, millions) iterations of the search process, one or more adjustments are then made to convert the current solution to a slightly different feasible solution in the local neighborhood of the solution space. As others have previously noted (e.g., Nissen & Paul, 1995; Marshall et al., 2003), a threshold accepting search process is similar to, and perhaps conceptually simpler than, a simulated annealing search process (Chap. 9).

For the following discussion, we will assume that there is no ability to utilize infeasible solutions during a search. Others have proposed methods to allow strategic oscillation of a search outside of the feasible region of the solution space by penalizing the infeasibilities, or by using methods that limit the amount of time spent exploring the infeasible area of the solution space (Wei et al., 2023). We will also assume in the following discussion that the search process addresses a mixed- or pure-integer problem, where the decision variables are assigned integer values (e.g., 0 = no, 1 = yes). Other forms of problems can be addressed with threshold accepting, such as those that were described in the linear programming chapter. In comparison to other search processes, a threshold accepting search process progresses in a somewhat convulsive yet controlled manner through the solution space in search of a high quality solution to a problem. Importantly, as with other heuristic search processes, a threshold accepting search process might be characterized as a combinatorial approach for solving problems that does not guarantee optimality will be achieved.

2 Background

In terms of the mechanisms that guide the search for the optimal solution to a problem, a threshold accepting search process might be considered a somewhat rational approach. A threshold accepting search process may begin with a feasible integer solution (forest plan) to a forest management problem, an assumption about the initial threshold value, and an assumption about how it will change as the current solution forest plan evolves. This heuristic then tracks the evolution of a single forest plan as it is changed through random modifications to the assignments of management actions to the stands. The threshold value is the amount of the decline in the objective function value (assuming a maximization objective) that is allowed when a forest harvest schedule is altered during the search. If a proposed change to the current solution results in a decrease in the objective function value, this change is only acceptable if the decline in value is smaller in magnitude than the current threshold value. The threshold value decreases as the number of iterations during the search progresses. The reduction in the threshold continues until it is so small that the search process becomes nearly a greedy, hill-climbing algorithm, at which point it terminates. This type of search process allows large deviations in objective function values in the initial stages of the search (diversifying the search), and small deviations in objective function values later in the search (intensifying the search).

The basic implementation of threshold accepting requires assumptions concerning several search-related parameters:

1. The beginning threshold level (in objective function units). Example: $100,000 or $1,000,000.
2. The ending threshold level (the stopping criteria). Example: $1.
3. The number of iterations allowed per threshold level. Example: 1, or 100, or 1000 iterations per threshold level. Perhaps a static amount of 100 iterations when the threshold is $100,000 (USD), 100 iterations when the threshold is $99,990, and so on.
4. The mechanism for changing the threshold.

3 Process

3.1 General Description

In developing a plan of action for the management of a forest, a basic process (Fig. 8.1) for a threshold accepting search process that seeks to maximize an objective may involve the following sequence of steps:

1. Develop an initial feasible integer solution to the problem (Step 1). This becomes both the current and the best solution to the problem, by default. The initial feasible solution can be developed randomly or deterministically or may simply be represented by a null (empty) set of management actions. For research purposes, a random initial solution is generally preferred.
2. While the threshold > some minimum value (using assumption 2 above).

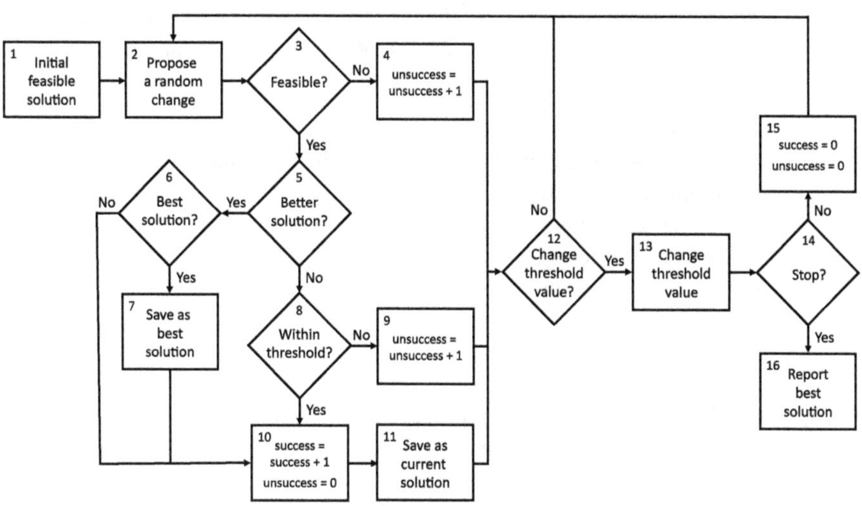

Fig. 8.1 A general threshold accepting search process. Credit: Pete Bettinger

- Propose a random change to (select a random neighbor of) the current solution, creating a temporary proposed solution (Step 2).
- Determine whether the temporary proposed solution is feasible with respect to all constraints (Step 3).
  - If the temporary proposed solution is infeasible, increase the number of consecutive infeasible solutions that have been proposed (Step 4).
  - If the temporary proposed solution is feasible, determine whether it will have an objective function value that is higher in quality than the current solution to the problem (Step 5).

    If the temporary proposed solution has an objective function value that is higher in quality than the current solution, determine whether this objective function value will be higher in quality than the objective function value of the best solution held in memory (Step 6). If so, save the temporary proposed solution as the best solution. In either event (yes or no), increment the successful changes at this threshold level, and reset the consecutive infeasible solutions to 0 (Step 10).

    If the temporary proposed solution does not have an objective function value that is higher in quality than the current solution, determine whether the difference is smaller than the current threshold (Step 8). If so, increment the successful changes at this threshold level, and reset the consecutive infeasible solutions to 0 (Step 10). Then save the temporary proposed solution as the current solution (Step 11). If not, reject the potential change, and increase the number of consecutive infeasible solutions that have been proposed (Step 9).

- Determine whether the threshold value needs to be changed (Step 12), and if not, return to Step 2. Otherwise, change the threshold value (Step 13).
- Given the new threshold value, determine whether to terminate the search process (Step 14).
  - If the threshold value is smaller than the stopping criteria, terminate the search process and report the best solution that was stored in the memory of the computer (Step 16).
  - If the threshold value is larger than the stopping criteria, reset the counter that tracks the successful feasible iterations at a threshold value, and the counter that tracks the consecutive infeasible iterations (Step 15). Then return to Step 2.

When a threshold accepting search process is employed upon a maximization problem (Fig. 8.2), one should expect to see the objective function value at each iteration of the model shift up and down (up preferably, but down when an inferior change is allowed (the decline in value is within the threshold). Here, the values should fluctuate within the range of Z, where Z = (best solution - threshold). When a

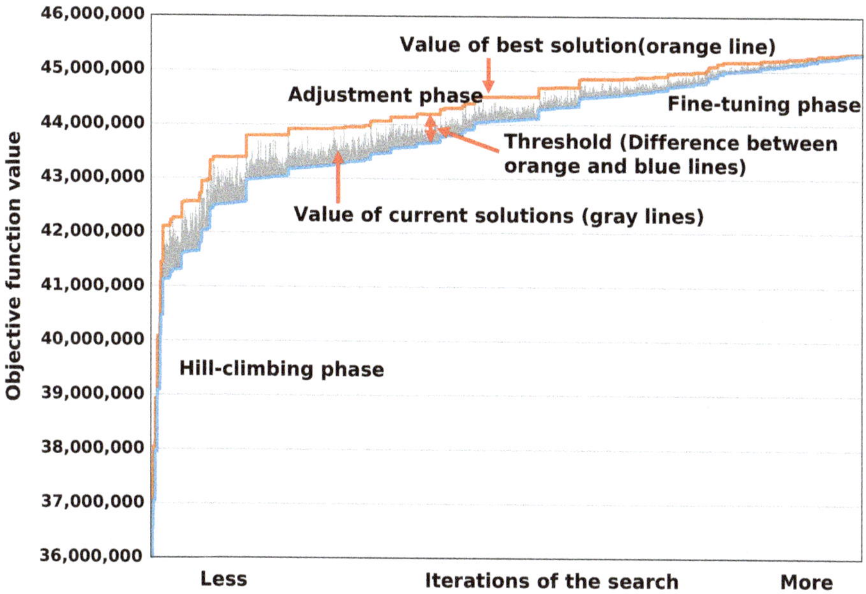

Fig. 8.2 Behavior of a threshold accepting search process when applied to a forest planning problem with a maximization objective. Credit: Pete Bettinger

threshold accepting search process is employed upon a minimization problem, (Fig. 8.3), one should expect to see the objective function value also fluctuate within the range of Z, where $Z =$ (best solution + threshold), ideally migrating downward toward the minimum possible value (perhaps 0) of the problem being addressed. Potentially, Z might employ the current objective function value rather than the best objective function value; however, doing so allows the search process to deviate widely from the best solution's objective function value.

3.2 Candidate Moves

A candidate move involves changing one or more characteristics of the current solution in order to create a different, temporary, candidate solution. This different, temporary, candidate solution is considered a "neighbor" in the solution space that is not too different than the current solution. Within a search process, *n*-opt moves are an enhancement mainly applied to *s*-metaheuristics that further refine the quality of the current solution and act to intensify the search around beneficial areas of the solution space. In general, a threshold accepting search process employs 1-opt moves.

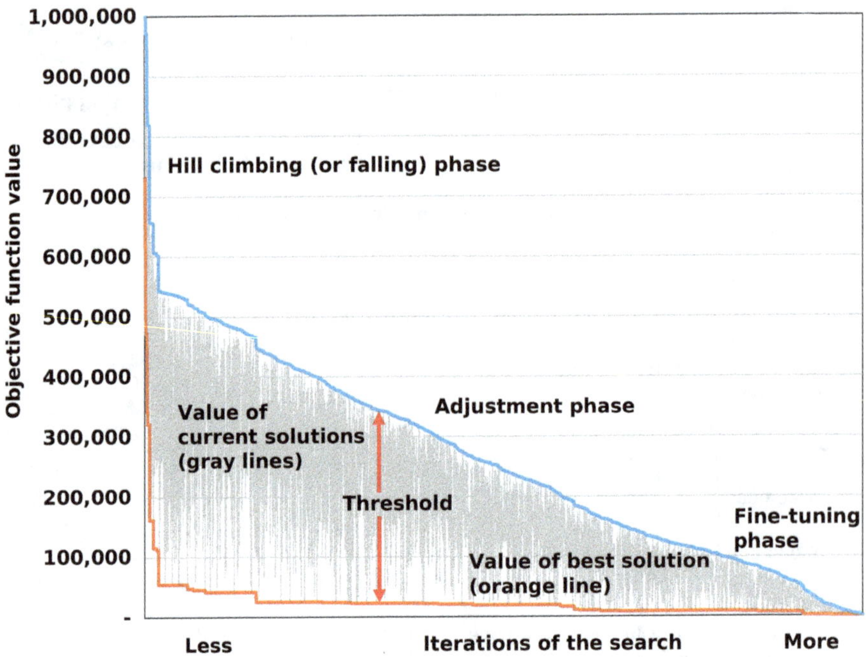

Fig. 8.3 Behavior of a threshold accepting search process when applied to a forest planning problem with a minimization objective. Credit: Pete Bettinger

> **Example**
>
> In employing 1-opt moves, a change is made to the state or quality of one of the decision variables in the problem. For example, with respect to forest harvest scheduling, the final harvest year of a stand of trees might be changed from the year 2026 to the year 2027. Alternatively, the management regime assigned to a stand might be changed from #3 (assigning a thinning in year 2025 and a final harvest in year 2035) to #27 (assigning a thinning in year 2025, another thinning in year 2032, and a final harvest in year 2040). As is evident, with 1-opt moves one decision variable (a timber stand) is selected, and a change is made to the assignment of management activity to that variable, whether the assignment within the harvest scheduling process consists of just one event (a final harvest) or a series of events (a management regime).

Others have previously described methods for utilizing 2-opt moves within a threshold accepting search process (Nissen & Paul, 1995; Tarantilis & Kiranoudis, 2002; Tarantilis et al., 2004). In these cases, two decision variables would be selected, and their assignments would change simultaneously during the harvest scheduling process. These types of changes to the current solution in a search process can also involve three (3-opt) or four (4-opt) or more decision variables (Glover, 1996). However, as more decision variables are involved in *n*-opt moves, the constraint assessments, particularly harvest adjacency constraints, may be more time-intensive. A typical 2-opt move would entail changing the state or quality of two of the decision variables simultaneously. This can be accomplished in one of two ways:

A. (more commonly) *Exchange* the assignment between two stands. For example, one stand might be scheduled for final harvest in year 2028 and a second stand might be scheduled for final harvest in the year 2031. A 2-opt exchange move would attempt to schedule the first stand in year 2031 and the second in year 2028, essentially switching the assignment among the two decision variables. The incorporation of 2-opt exchange moves has been shown to significantly improve the quality of solutions developed by *s*-metaheuristics (Bettinger et al., 1999; Bettinger & Boston, 2017).
B. (less commonly) Simultaneously and independently change the assignments of two stands. For example, one stand might be scheduled for final harvest in year 2028 and a second stand might be scheduled for final harvest in the year 2031. A 2-opt change move might attempt to schedule the first stand in year 2030 and the second in year 2035, essentially changing the assignment among the two decision variables. At first glance, this type of 2-opt move may simply represent two successive 1-opt moves. However, there is a chance that this type of simultaneous change to decision variables in the search process can lead to higher quality solutions, because the impact on the objective function value of both changes made simultaneously may be less severe than when they are made sequentially.

The exchange type of 2-opt moves cannot be made in isolation from 1-opt moves (or 2-opt change moves). With the use of 2-opt exchange moves, no adjustment to the number of assignments (e.g., there are ten final harvests scheduled during the first year of a forest plan, and eight in the second year, and so on) can be made. There will always be the same number of each type of assignment, since they are simply switched from one decision variable to the next. Therefore, it has been suggested that a search process vacillate back and forth between 1-opt and 2-opt (or higher) moves to allow diversity of assignment into the search process (Bettinger et al., 2015). Alternatively, the end of a 1-opt search by an *s*-metaheuristic may be complemented with some effort applied to the use of 2-opt moves (to refine the solutions). This is not an issue with 2-opt (or higher) change moves. However, in one application of 2-opt and higher change moves within a forest planning *s*-metaheuristic, it was suggested that the performance of the search process might not be enhanced when two or more decisions are changed independently and simultaneously (Bachmatiuk et al., 2015).

Example

The value of a 2-opt move over a 1-opt move is illustrated in the following example. Imagine that in each year of a forest plan, several stands of trees are scheduled for harvest (Fig. 8.4). Further, assume that the goal is to schedule an even amount of harvest volume over time. A 1-opt search process would change the state or harvest year of one stand at a time. This change may result in a condition that is unacceptable with respect to the parameters of the search (a new solution that is too far away from the previous solution in terms of its objective function value). For example, the amount of volume scheduled for harvest during time period 1 of iteration 2 in Fig. 8.4 may deviate too far from the scheduled volumes in other time periods, and this potential change to the harvest schedule may be unacceptable. Thus, getting to a better place (the harvest schedule represented in iteration 3) in the solution space may be impossible. However, a 2-opt search process might simultaneously make the two changes (Fig. 8.5) before an assessment of the resulting objective function value occurs, thus perhaps allowing the search to locate better solutions to the problem being solved.

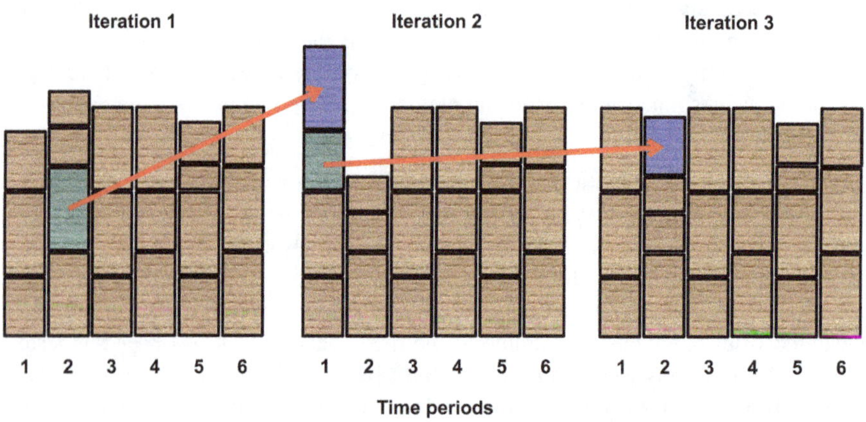

Fig. 8.4 The general effect on wood flow using 1-opt moves within a heuristic. Volume from separate stands of trees scheduled for harvest during each time period is represented by the individual blocks. For example, during time period 1 of iteration 1, three stands are scheduled for harvest, each having different amounts of volume. Credit: Pete Bettinger

3.3 *Threshold Value*

The use of a firm threshold guarantees that a move away (in the solution space) from the current solution to a proposed, temporary candidate solution will only be successful (made) if the objective function value of the proposed, temporary

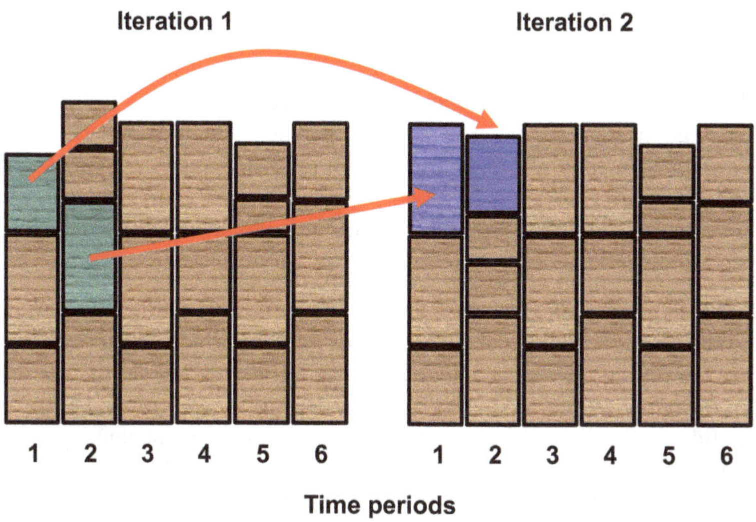

Fig. 8.5 The general effect on wood flow using 2-opt moves within a heuristic. Credit: Pete Bettinger

candidate solution is not much worse in quality than the objective function value of the best (or previous) solution (Nurmela, 1993). Over the course of a threshold accepting search process, the threshold value will initially be a relatively large value and decrease in size until it becomes too small to continue the search. If the threshold value reaches nearly zero (0) the search process becomes a stochastic hill-climbing process. The initial threshold value represents a relatively large upper (or lower, depending on the problem being solved) bound on the objective function value of the current (or best) solution (Lin et al., 1995), whereby the search is allowed to wander through the solution space, contacting or locating feasible solutions that have objective function values within this bound. An adaptive threshold value that is based on search performance (time since last improvement in the value of the best solution) may also be employed in cases where the solution space is particularly challenging for *s*-metaheuristic search processes to escape local optima (Lin et al., 1995). These adjustments may facilitate a longer search by increasing the threshold value when it is observed that the search may have become limited to an area around a local minimum in the solution space.

3.4 *Threshold Change*

A change in the threshold (or the *decrement rule* (Lin et al., 1995) can be linear (constant), geometric (reduced by a factor), or otherwise mathematical in nature. General methods Talbi (2009) suggested for changing the temperature of a simulated annealing search process (Chap. 9) may also apply to a threshold accepting search process:

- Reduce the threshold by a constant amount (linear reduction).
- Reduce the threshold by a constant factor (geometric reduction).
- Reduce the threshold by a logarithmic factor (new threshold = old threshold/Log [search iterations]).
- Reduce or increase the threshold by non-monotonic or adaptive factors that depend on search history.

Dueck and Scheuer (1990) suggest that results from using a threshold accepting search process may be insensitive to the manner in which the threshold value changes. However, one might utilize a geometric rate (or constant factor such as 0.999 of the previous threshold value) to effectively allow the heuristic to spend less working time in the diversification phase of the search (in the beginning) and more time working in the intensification phase of the search (near the end). In general, a threshold might change after some assumed number of *successful* iterations of the search process. These are iterations where a move is suggested, and the result is a feasible solution whose objective function value satisfied the threshold accepting criteria. However, a threshold accepting search process (along with a simulated annealing search process) can be saddled by issues related to a proposed *Principle of Evolving Standards*:

> *As the criteria within a heuristic for accepting inferior, feasible solutions tighten, more potential moves (changes to the current solution) are rejected, resulting in an extended time to termination of the search process.*

To counteract the increasing number of unsuccessful changes to the current solution as the threshold value contracts, two possible options might be employed within the heuristic:

A. Count every attempt that is made to change the current solution, whether feasible or infeasible. Then, change the threshold value after every *n* attempts have been made. On a positive note, this strategy generally results in a shorter computational time required to complete the heuristic search process. On the other hand, there may be a very high proportion of infeasible potential moves to feasible potential moves. And in these cases, no change to a current solution may be made when the threshold levels are very small, approaching the minimum level prior to termination of the process. This defeats the purpose of search intensification.

```
  Do while threshold > 1
  ...
  Whether feasible or not, count = count + 1
  If (count = time to change threshold) then
    change threshold
    count = 0
  endif
Loop
```

B. Count every consecutive infeasible or unsuccessful (due to the proposed solution having an objective function value too far away from the current [or best] solution) attempt that is made to change the current solution and reset the count to 0 when a feasible change is made to the current solution. Then after *n* consecutive infeasible solutions have occurred, the threshold value is changed. This strategy increases the computational time required to complete the heuristic search process but promotes intensification at the end of the search process by not allowing the search to terminate so quickly.

```
 Do while threshold > 1
 ...
 If the attempted change is infeasible or unsuccessful, then
   count = count + 1
 else if the attempted change is feasible, then
   count = 0
 endif
 If (count = time to change threshold) then
   change threshold
   count = 0
 endif
Loop
```

This latter strategy may be one of the most important adjustments made to a standard threshold accepting search process. The methods for utilizing an unsuccessful move strategy were described nearly three decades ago (Siedentopf, 1995). Many of the forestry-related works described in Sect. 4 have employed this strategy. For example, Coulter et al. (2006) also used the unsuccessful count as a way of terminating the algorithm.

3.5 Search Reversion

Search reversion is an interesting concept in that it controls the amount diversification that is allowed while also promoting intensification around high quality solutions. Reversion was listed as a search diversification strategy in Thomas and Salhi (1998), but often it behaves more like an intensification strategy.

Example

Imagine that a search process has found a very good solution to a problem (let us call it Solution A). Then, the search process looks for higher quality feasible solutions in the nearby neighborhood, along the Markov chain (Frausto-Solis et al., 2021). The Markov property suggests that future states of a system are conditionally independent of past states, and that additional knowledge about the past state is of no value in making future decisions (Clarke & Disney, 1985). So perhaps the search process randomly locates feasible, but lower quality Solutions B, C, D, etc. If a higher quality feasible solution (Solution M) is eventually located, then it might replace the best solution stored in computer memory (Solution M replaces Solution A as the best solution). However, if after some time of searching (stochastically or deterministically, depending on the heuristic), a higher quality feasible solution has not been found, the search stops, the current solution is replaced by (or reverts back to) Solution A, and the search process begins again from there.

Unless a search process has a stochastic element embedded within it, search reversion will be useless, as a deterministic process will likely revert continuously to the same set of solutions. A very simple reversion strategy for threshold accepting might involve the following:

```
time_to_revert = 10
Do while threshold > 1
  ...
  revert_count = revert_count + 1
  If (revert_count = time_to_revert) then
    replace the current solution with the best solution in memory
    revert_count = 0
  endif
  ...
Loop
```

In the example above, the search process will pause every ten iterations, and the current solution (whatever it looks like) will be replaced with the characteristics or assignments that represent the best solution (whatever that looks like) stored in memory.

The rate of reversion, or number of iterations that pass before the best solution replaces the current solution, is a parameter of the search process that needs investigation. The appropriate reversion rate seems to differ from problem to problem, but in tests of forestry problems, it has been typically less than 50 feasible iterations of a search process (Bettinger & Zhu, 2006; Bettinger et al., 2015). A reversion rate of ten iterations or less does not allow a significant amount of diversification before restarting the process with the best solution that was stored away in computer memory. However, when compared to search processes that employ no reversion, the use of search reversion usually results in higher quality solutions. No reversion essentially assumes a reversion rate of 0 if the code for the process noted above has been incorporated into a heuristic. Long reversion rates (100+ feasible iterations of a search) also generally do not necessarily lead to higher quality solutions, likely because intensification around high quality solutions is limited.

3.6 Termination Rules

One of the main issues with a threshold accepting search process is in designing how the heuristic will terminate or stop. Theoretically, the search process stops when the threshold declines to some minimal level close to zero in value. For example, if the objection function of a forest plan was measured in dollars or Euros, the process might terminate when the threshold drops below one dollar or one Euro. Alternatively, a heuristic search process might terminate after a pre-determined number of iterations have passed where no improvement to the quality of the best overall solution has been noticed (Lin et al., 1995). This type of termination tactic is meant to prevent wasting time, but careful thought needs to be applied toward the decision regarding how much time (or iterations) has passed since the last improvement in the quality of the best solution. Early termination may occur with the time (or iterations) allowed is short (or small). Another tactic would simply require a pre-defined number of consecutive unsuccessful (or infeasible) iterations to pass, triggering the termination of the search process. This tactic pays no attention to the threshold level, and unfortunately may result in early termination of the search process if the number of consecutive unsuccessful (or infeasible) iterations that are allowed is limited.

3.7 Optimizing the Parameters

Through the application of threshold accepting to numerous forest planning problems, it has become evident that a universal set of search parameters is elusive. As others have suggested (Perea et al., 2008), the use of a threshold accepting search process may require calibration of a number of parameters (described above) whose values can vary based on the problem being addressed. Often, trial-and-error processes are employed, using reasonable ranges of parameter values, to arrive at a set that might be used to effectively solve a problem. In their seminal work, Pukkala and Heinonen (2006) suggested that the optimal parameters for a threshold accepting search process might be determined by employing another independent search process (Hook and Jeeves direct search approach) to explore the impact of the parameters on the fitness (objective function value) of the resulting solutions. Jin et al. (2016) further suggested the use of procedures such as these to help determine the best set of parameters needed for heuristic search processes.

Selecting the appropriate parameters for a heuristic such as threshold accepting is a pervasive problem of all heuristics that are designed to escape and break free of local optima during a search process (Michalewicz & Fogel, 2002). Selecting the appropriate parameters through trial-and-error is time-consuming and often ignores the interaction effects among them, as multiple parameters may not act independently on modifying the behavior of a search process. Further, the process of *parameter tuning* a heuristic is often conducted before a search process begins. Parameter adjustment processes that occur during a search process (whether adaptive to the conditions of a search, or deterministically defined) effectively remain an open area of work with respect to the use of operations research methods in forestry and natural resource management.

4 Contemporary Forest Harvest Scheduling Problem

As a threshold accepting search process may require thousands or millions of iterations prior to terminating, only one iteration of the process is described here. Assume that a number of assignments of final harvest time periods to timber stands have been made for the southern Arkansas forest described in Sect. 4 of Chap. 2. Perhaps these assignments are represented by the random search solution described in Table 7.1 from Chap. 7 of this book. In this solution, the objective function value was $4,023,263.55, and for this example assume it represents the best solution located thus far during the search process. Further assume that the current threshold value in a threshold accepting search process is $5000. Within the search process, stand 17 and final harvest period 8 might both be randomly selected as the potential move to make in the local neighborhood of solutions (from stand 17 and final harvest period 7). The objective function value contribution from scheduling stand 17 for a final harvest during time period 7 was $138,124.96. The objective function value contribution from potentially scheduling stand 17 for a final harvest during time period 8 is $133,573.88. Making this random move through the solution space could result in a decline in objective function value of $4551.08 ($138,124.96 - $133,573.88). Since the decline in objective function value is smaller than the current threshold value, the proposed change in the harvest schedule is acceptable. This assumes, of course, that none of the constraints will be violated when this change to the harvest schedule is made. Had the threshold value been smaller than $4551.08, this particular potential change in the assignment of a final harvest time period to stand 17 would not have been permitted.

5 Applications

Along with works pertaining to other domains of commerce (e.g., Tarantilis et al., 2004; Perea et al., 2008), methods for using threshold accepting search processes to address forest planning problems have been described (Bettinger et al., 2002, 2003, 2015; Zhu et al., 2007; Zhu & Bettinger, 2008; Li et al., 2010; Akbulut et al., 2017). One example of using threshold accepting in a forest planning context involved the scheduling of harvests while maintaining or developing northern spotted owl (*Strix occidentalis caurina*) habitat conditions in western Oregon (Bettinger & Boston, 2008). This endeavor had a 40 year time horizon that was divided into eight time periods, each 5 years long. The objective function sought to minimize the deviations between scheduled harvest volumes and a target harvest volume for the entire landscape during each time period, subject to harvest adjacency and green-up, minimum harvest age, and minimum habitat constraints. The landscape to which the process was applied was about 580,685 acres (about 235,000 ha) and contained about 39,500 management units (stands). The threshold accepting process was developed to accommodate a Model I problem where timber harvests were potentially scheduled for each management unit during the time horizon. In this implementation of threshold accepting, only 1-opt candidate moves were accommodated within the heuristic. The stopping criteria included the deterioration of the threshold below some minimum value, as well as a limit on the number of consecutive infeasible or unsuccessful moves, and a limit on the number of consecutive iterations without an improvement in the best overall solution. Different minimum harvest age, minimum habitat level, and clearcut size restrictions were modeled to assess the impacts of these potential policies on forest management outcomes within the landscape that was studied.

6 Summary

Threshold accepting is very similar to other *s*-metaheuristic search processes, whereby a single solution to a problem is temporarily changed through the employment of 1-opt, 2-opt or *n*-opt moves. Generally speaking, if these adjustments to the solution lead to a higher quality objective function value, and the resulting solution is feasible with respect to all constraints, the adjustments are accepted, and the solution is changed. When the adjustments lead to a lower quality objective function value while maintaining feasibility, the adjustments may be accepted as long as the decline in objective function value is within a threshold value. This manner of accepting lower quality solutions during the search process is what differentiates threshold accepting from simulated annealing and the demon algorithm (Creutz, 1983; Wood & Downs, 1998). This manner of accepting lower quality solutions during the search is also conceptually simpler than the other search processes. Further, the effectiveness of a threshold accepting search process for locating high quality solutions to forest harvest scheduling problems has been found in several instances to be as good as the effectiveness of a simulated annealing search process.

Take Home Message

Threshold accepting is a search process that enhances a stochastic hill-climbing search process by accepting inferior moves through the solution space while searching for the optimal solution, as long the resulting solutions are not too much worse than the previous (or best) solution held in memory. When an inferior move is proposed, a decision is made to either accept the move or reject the move based simply on how far away (in objective function value units) it is from either the current solution or the best solution held in memory. Early in the search, when the threshold is large, search diversification is evident. Later in the search, when the threshold is small, search intensification is evident. This ability to accept inferior moves helps the search process avoid becoming trapped at local optima, a common problem with hill-climbing search processes. This ability allows the search process to move to lower quality solutions with the hope of eventually locating other higher quality solutions.

Questions

1. If unsuccessful attempts to change the current solution during a threshold accepting search were not recognized, how might the search process behave?
2. Toward the end of the fine-tuning phase of a threshold accepting search process, how does the search process generally behave?
3. What might be the value of a 2-opt change move during a threshold accepting search process, and would there be a need for 1-opt moves if it were employed?
4. For contemporary forest harvest scheduling problems, what are some disadvantages of using 4-opt moves during a threshold accepting search process?
5. If the initial value of a threshold accepting search process was very large with respect to the objective function value of a feasible solution, how does the search process behave during the hill-climbing phase?
6. If the initial value of a threshold accepting search process was very small with respect to the objective function value of a feasible solution, how does the search process behave during the hill-climbing phase?
7. In the basic implementation of threshold accepting, the threshold value decreases from the initial assumed value as the search progresses to a very small (or zero) value at the termination of the search. Why do we not assume that the threshold value can be negative?

References

Akbulut, R., Bettinger, P., Ucar, Z., Obata, S., Boston, K., & Siry, J. (2017). Spatial forest plan development using heuristic processes seeded with a relaxed linear programming solution. *Forest Science, 63*(5), 518–528.

Albus, J. S. (1981). *Brains, behavior, and robotics*. BYTE Publications, Inc.

Bachmatiuk, J., Garcia-Gonzalo, J., & Borges, J. G. (2015). Analysis of the performance of different implementations of a heuristic method to optimize forest harvest scheduling. *Silva Fennica, 49*(4), Article ID 1326.

Bettinger, P., & Boston, K. (2008). Habitat and commodity production trade-offs in Coastal Oregon. *Socio-Economic Planning Sciences, 42*(2), 112–128.

Bettinger, P., & Boston, K. (2017). Forest planning heuristics—current recommendations and research opportunities for *s*-metaheuristics. *Forests, 8*(12), Article 476.

Bettinger, P., Boston, K., & Sessions, J. (1999). Intensifying a heuristic forest harvest scheduling search procedure with 2-opt decision choices. *Canadian Journal of Forest Research, 29*(11), 1784–1792.

Bettinger, P., Demirci, M., & Boston, K. (2015). Search reversion within *s*-metaheuristics: Impacts illustrated with a forest planning problem. *Silva Fennica, 49*(2), Article ID 1232.

Bettinger, P., Graetz, D., Boston, K., Sessions, J., & Chung, W. (2002). Eight heuristic planning techniques applied to three increasingly difficult wildlife planning problems. *Silva Fennica, 36*(2), 561–584.

Bettinger, P., Johnson, D. L., & Johnson, K. N. (2003). Spatial forest plan development with ecological and economic goals. *Ecological Modelling, 169*(2–3), 215–236.

Bettinger, P., & Zhu, J. (2006). A new heuristic for solving spatially constrained forest planning problems based on mitigation of infeasibilities radiating outward from a forced choice. *Silva Fennica, 40*(2), 315–333.

Clarke, A. B., & Disney, R. L. (1985). *Probability and random processes: A first course with applications* (2nd ed.). John Wiley & Sons, Inc.

Coulter, E. D., Sessions, J., & Wing, M. G. (2006). Scheduling forest road maintenance using the analytic hierarchy process and heuristics. *Silva Fennica, 40*(1), 143–160.

Creutz, M. (1983). Microcanonical Monte Carlo simulation. *Physical Review Letters, 50*(19), 1411–1414.

Dueck, G., & Scheuer, T. (1990). Threshold accepting: A general purpose optimization algorithm appearing superior to simulated annealing. *Journal of Computational Physics, 90*(1), 161–175.

Frausto-Solis, J., Hernández-Ramírez, L., Castilla-Valdez, G., González-Barbosa, J. J., & Sánchez-Hernández, J. P. (2021). Chaotic multi-objective simulated annealing and threshold accepting for job shop scheduling problem. *Mathematical and Computational Applications, 26*(1), Article 8.

Glover, F. (1996). Finding a best travelling salesman 4-opt move in the same time as a best 2-opt move. *Journal of Heuristics, 2*(2), 169–179.

Jin, X., Pukkala, T., & Li, F. (2016). Fine-tuning heuristic methods for combinatorial optimization in forest planning. *European Journal of Forest Research, 135*(4), 765–779.

Li, R., Bettinger, P., & Boston, K. (2010). Informed development of meta heuristics for spatial forest planning problems. *The Open Operational Research Journal, 4*, 1–11.

Lin, C. K. Y., Haley, K. B., & Sparks, C. (1995). A comparative study of both standard and adaptive versions of threshold accepting and simulated annealing algorithms in three scheduling problems. *European Journal of Operational Research, 83*(2), 330–346.

Marshall, H., Boston, K., & Sessions, J. (2003). Harvest scheduling with aggregation adjacent constraints: Threshold acceptance approach. In D. Briggs (Ed.), (Chair) *Precision forestry, proceedings of the second international precision forestry symposium* (pp. 131–135). University of Washington, College of Forest Resources, Institute of Forest Resources.

Michalewicz, Z., & Fogel, D. B. (2002). *How to solve it: Modern heuristics*. Springer-Verlag.

Nissen, V., & Paul, H. (1995). A modification of threshold accepting and its application to the quadratic assignment problem. *Operations Research Spektrum, 17*(2–3), 205–210.

Nurmela, K. J. (1993). *Constructing combinatorial designs by local search*. (Series A: Research Reports. No. 27). Helsinki University of Technology, Digital Systems Laboratory.

Perea, C., Alcala, J., Yepes, V., Gonzalez-Vidosa, F., & Hospitaler, A. (2008). Design of concrete bridge frames by heuristic optimization. *Advances in Engineering Software, 39*(8), 676–688.

Pukkala, T., & Heinonen, T. (2006). Optimizing heuristic search in forest planning. *Nonlinear Analysis: Real World Applications, 7*(5), 1284–1297.

Siedentopf, J. (1995). *An efficient scheduling algorithm based on threshold accepting*. (Report No. 16). University of Leipzig, Institute of Production Management and Industrial Information Management.

Talbi, E.-G. (2009). *Metaheuristics. From design to implementation*. John Wiley & Sons, Inc.

Tarantilis, C. D., & Kiranoudis, C. T. (2002). A threshold-accepting metaheuristic method for scheduling the operations of dehydration plants. *Drying Technology, 20*(6), 1143–1160.

Tarantilis, C. D., Kiranoudis, C. T., & Vassiliadis, V. S. (2004). A threshold accepting metaheuristic for the heterogeneous fixed fleet vehicle routing problem. *European Journal of Operational Research, 152*(1), 148–158.

Thomas, P. R., & Salhi, S. (1998). A tabu search approach for the resource constrained project scheduling problem. *Journal of Heuristics, 4*(2), 123–139.

Wei, Z., Hao, J. K., Ren, J., & Glover, F. (2023). Responsive strategic oscillation for solving the disjunctively constrained knapsack problem. *European Journal of Operational Research, 309*(3), 993–1009.

Winkler, P., & Fang, K. T. (1997). Application of threshold-accepting to the evaluation of the discrepancy of a set of points. *SIAM Journal on Numerical Analysis, 34*(5), 2028–2042.

Wood, I., & Downs, T. (1998). Demon algorithms and their application to optimization problems. In P. K. Simpson (Ed.), (General chair) *IEEE World Congress on Computational Intelligence* (pp. 1661–1666). Institute of Electrical and Electronics Engineers, Inc.

Zhu, J., & Bettinger, P. (2008). Estimating the effects of adjacency and green-up constraints on landowners of different sizes and spatial arrangements located in the southeastern U.S. *Forest Policy and Economics, 10*(5), 295–302.

Zhu, J., Bettinger, P., & Li, R. (2007). Additional insight into the performance of a new heuristic for solving spatially constrained forest planning problems. *Silva Fennica, 41*(4), 687–698.

Chapter 9
Simulated Annealing Search

Learning Objectives
The objectives of this chapter on simulated annealing are to help the reader

- Understand the history and motivation of the simulated annealing search process
- Understand the general manner in which simulated annealing operates
- Understand the use and value of limited diversification in local search options
- Understand the main limitations associated with the use of simulated annealing

1 Introduction

Simulated annealing is one of the oldest and most widely used and studied heuristic search process (Franzin & Stützle, 2019). Technically, it is an *s*-metaheuristic search process that follows the development and evolution of a single solution to a problem (a feasible forest harvest schedule) along a path that hopefully converges at the global optimum solution to a problem, although only approximate, near-optimal solutions should be expected (Černý, 1985). In this sense, it is a local search process where one potential solution is examined with each iteration of the model, and therefore a single feasible solution (the *current* feasible solution) serves as the basis for a potential move to a neighboring solution in the solution space (Nurmela, 1993). Simulated annealing has also been called *Monte Carlo annealing*, *probabilistic exchange*, *stochastic relaxation*, and *statistical cooling* (Michalewicz & Fogel, 2002). Glover (1986) considered it a *random shakeup approach* for solving complex problems, a label that might be applied equally as well to threshold accepting. As a search process employed for solving complex problems, it can be characterized as a point-based or *s*-metaheuristic (Talbi, 2009), and like threshold accepting, it functions as a unidirectional approximation algorithm (Siedentopf, 1995), a refined local search algorithm (Winkler & Fang, 1997), and a convergent algorithm (Lin et al., 1995).

P. Bettinger, *Forest Harvest Scheduling*, Springer Texts in Business and Economics,
https://doi.org/10.1007/978-3-031-89432-9_9

The simulated annealing search process was motivated by basic concepts from the physical sciences, where the temperature of a heated material when separated from a heat source will cool down (decline) to a thermal equilibrium (Glover & Greenberg, 1989). The Metropolis algorithm forms the basis for simulated annealing, and its introduction through a 1953 publication (Metropolis et al., 1953) signaled the beginning of the use of Monte Carlo methods for solving complex problems (Gubernatis, 2005). Interestingly, the authors of the paper were scientists and mathematicians associated with the Manhattan Project (Edward Teller perhaps the most famous), and none of them continued work in this area after the paper was published (Gubernatis, 2005). Central to the algorithm is the use of a function from the field of statistical mechanics that is very similar to the Boltzmann probability factor, which helps describe the distribution of energy among atoms (Broda & Gay, 1983; Kirkpatrick et al., 1983). Within a simulated annealing search process, this relationship assists in efforts to determine whether to transition from a current, feasible solution to another feasible, yet inferior solution. An important note to make here is that like a greedy algorithm, a search process using the Metropolis algorithm concept moves from one solution to another (making it an *s*-metaheuristic). Unlike a greedy algorithm, a search process using the Metropolis algorithm concept possibly allows inferior moves to be made within the solution space, as changes to the current solution are allowed based on a transition probability. Like threshold accepting, simulated annealing progresses in a somewhat convulsive yet controlled manner through the solution space in search of a high quality solution to a problem. However, unlike threshold accepting, there is generally no firm threshold to prevent making a change to the current solution when the change results in a solution with a far inferior objective function value (Nurmela, 1993).

As a search process, simulated annealing is analogous to one aspect of thermodynamics—the manner in which metals or glasses cool down from a very hot mass and form a crystallized structure (Černý, 1985; Nurmela, 1993). Annealing is the process of slowly cooling a material whose initial temperature was high, strengthening it and possibly removing imperfections. Therefore, beginning from some high temperature state, as the temperature of a material is slowly lowered, annealing helps the material attain a thermal equilibrium state with maximum entropy and form a rigid structure (Černý, 1985; Brooks & Morgan, 1995). The rate of cooling is important in developing the structure, and in locating the highest quality end state. From a theoretical point of view, the process involves the current state (i) of a solid, a subsequent, temporary, state (j) that can be created through a small perturbation (adjustment) of i, and the energy states (E_i and E_j, the objective function values) of each (Nurmela, 1993). If E_j is higher in quality than E_i, j becomes the current state of the solid. If E_j is lower in quality than E_i, j may also become the current state of the solid based on a probability that is determined using on the difference between E_i and E_j and the progress of the search process (as represented by the current temperature of the system). In theory, if the temperature of the solid declines slowly during the process, and if there are a sufficient number of perturbations (adjustments) attempted at each temperature level, a thermal equilibrium might be reached at each temperature level (Nurmela, 1993).

Metropolis (1987) described the theoretical development of the simulated annealing search process as ...

> ... *if a statistical "move" of a particle in the system resulted in a decrease in the energy of the system, the new configuration was accepted. On the other hand, if there was an increase in energy, the new configuration was accepted only if it survived a game of chance biased by a Boltzmann factor.*

Hence, a search process employing simulated annealing allows lower-valued, inferior configurations of forest harvest schedules to be developed as long as they are able to pass a test. The test relies on two factors: the amount of the decrease in value that is expected by making a change from one feasible solution to another, and the amount of time that the process has been searching for the optimal solution. Theoretically, when cooling occurs too quickly, the structure of the metal is compromised. Similarly in problem-solving, when cooling occurs too quickly, the search process can become stranded at a local optimum very quickly (Michalewicz & Fogel, 2002). However, unlike human behavior, where upon reaching a destination there nearly always is a chance of moving elsewhere, simulated annealing utilizes a progressively vanishing probability of moving away from a local optimum, and because of this random behavior, it has been categorized as not very intelligent in conducting a search (Glover, 1986). Nonetheless, amongst the various applications one can find in the literature that describe the use of heuristic search to address forest harvest scheduling problems, a simulated annealing search process often emerges as one of the most efficient and effective for complex problems.

2 Background

In theory, a generalized Boltzmann distribution can be used to estimate the probability that a system (problem) will have a certain state (proposed solution), given the value of the current and proposed state of the system (Uffink, 2014). Again, in theory, a generalized Boltzmann distribution can be employed to estimate the probability of moving along a *Markov chain* from one solution to the next (within a local neighborhood) when solving a problem using an *s*-metaheuristic method (Černý, 1985). For moves throughout a solution space that are non-improving, the probability of moving from one state (current solution) to another (inferior to the current solution) can be based on the energy of the state (difference between proposed new state and either the current state or the best state saved in memory) and the temperature (an analogy for amount of effort already employed in solving the problem) of the system (Metropolis et al., 1953). States with lower energy (small difference between proposed new state and either the current state or the best state saved in memory) will have a high probability of being moved to, while states with high energy (large difference) will have a lower probability. Generally, during the beginning of a search for the optimal solution to a problem, the probabilities are high of moving from one state (the current solution) to another inferior state given the low amount of effort already employed in solving the problem. Later in the search, when temperatures are lower (when the amount of effort already employed in solving the problem is high), the probability of moving similarly one state (the current solution) to another inferior state will be much lower. In effect, the use of the Boltzmann factor in helping to decide whether to accept an inferior move along the Markov chain facilitates search diversification in the early stages and search intensification in the later stages.

With respect to forest harvest scheduling, a Markov chain represents the series of forest harvest schedules that are developed sequentially through local moves (small changes) in harvest timing or management regime allocation. In some cases, the sequence of moves from one state (forest harvest schedule) to the next state (a different, but superior forest harvest schedule) is made with 100% certainty. In other cases, the sequence of moves from one state (forest harvest schedule) to the next state (a different, but inferior forest harvest schedule) is based on a probability that depends on the time already invested in solving the problem and the difference in objective function values between the current state and the proposed state. As an alternative, this transition probability may depend on the difference between the best overall solution (quality of the best forest harvest schedule) found during the search and the proposed state.

3 Process

3.1 General Description

In developing a plan of action for the management of a forest, a basic process (Fig. 9.1) for a simulated annealing search process that seeks to maximize an objective may involve the following steps:

1. Develop an initial feasible integer solution to the problem (Step 1). This becomes both the current and the best solution to the problem, by default. The initial feasible solution can be developed randomly or deterministically or may simply be represented by a null (empty) set of management actions. For research purposes, a random initial solution is generally preferred.
2. While the temperature > some minimum value (using assumption 2 above) ...
   (a) Propose a random change to (select a random neighbor of) the current solution, creating a temporary proposed solution (Step 2).
   (b) Determine whether the temporary proposed solution is feasible with respect to all constraints (Step 3).

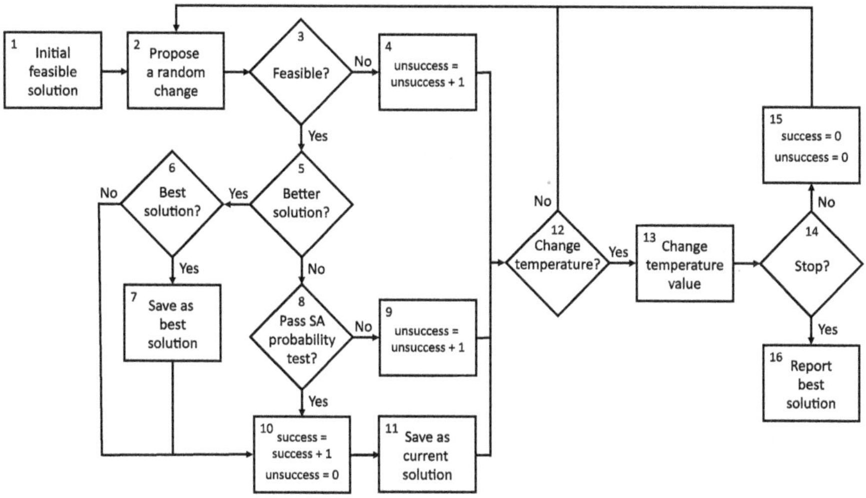

Fig. 9.1 A general simulated annealing search process. Credit: Pete Bettinger

- If the temporary proposed solution is infeasible, increase the number of consecutive infeasible solutions that have been proposed (Step 4).
- If the temporary proposed solution is feasible, determine whether it will have an objective function value that is higher in quality than the current solution to the problem (Step 5).
  - If the temporary proposed solution has an objective function value that is higher in quality than the current solution, determine whether this objective function value will be higher in quality than the objective function value of the best solution held in memory (Step 6). If so, save the temporary proposed solution as the best solution. In either event (yes or no), increment the successful changes at this threshold level, and reset the consecutive infeasible solutions to 0 (Step 10). Then save the temporary proposed solution as the current solution (Step 11).
  - If the temporary proposed solution does not have an objective function value that is higher in quality than the current solution, determine the probability of accepting this proposed change (Step 8).

$$\text{Probability} = \text{Exp}(-((\text{current solution} - \text{proposed solution})/\text{temperature}))$$

where
current solution = the objective function value of the current solution
proposed solution = the objective function value of the proposed solution

(a) Alternatively, the objective function value of the best solution may be used in lieu of the objective function value of the current solution.

(b) For minimization problems, Exp(– ((proposed solution – current solution)/temperature)) is used.

If the probability of accepting the inferior solution (which is presented in the range of 0–1) is larger than a randomly drawn number between 0 and 1, increment the successful changes at this threshold level, and reset the consecutive infeasible solutions to 0 (Step 10). Then save the temporary proposed solution as the current solution (Step 11). If not, reject the potential change, and increase the number of consecutive infeasible solutions that have been proposed (Step 9).

(c) Determine whether the temperature needs to be changed (Step 12), and if not, return to Step 2. Otherwise, change the temperature (Step 13). Given the new temperature, determine whether to terminate the search process (Step 14).

- If the temperature is smaller than the stopping criteria, terminate the search process and report the best solution that was stored in the memory of the computer (Step 16).
- If the temperature is larger than the stopping criteria, reset the counter that tracks the successful feasible iterations at a temperature value, and the counter that tracks the consecutive infeasible iterations (Step 15). Then return to Step 2.

3.2 *Candidate Moves*

Since simulated annealing is an *s*-metaheuristic search process, a candidate move involves changing one or more characteristics of the current solution in order to create a new, different, yet temporary candidate solution. This different, temporary, candidate solution is considered a "neighbor" in the solution space that is not too different in character than the assignments contained in the current solution. By default, simulated annealing employs 1-opt candidate moves, where a change is proposed to the state or quality of one of the decision variables. For example, the assigned final harvest (clearcut) year of a stand of trees might be changed from the year 2032 to the year 2027. As was noted in the chapter on threshold accepting, 2-opt moves (or *n*-opt moves) may enhance the quality of results obtained, yet the computational cost may be expensive when complex constraints need to be assessed. The use of 2-opt exchange of assignments within a simulated annealing search process has also been employed in other fields to locate high quality solutions to complex problems (e.g., Lee, 2001).

Like threshold accepting, if a simulated annealing search process allows more than one potential change to the current solution at each temperature level, these explorations have been suggested as occurring in the *inner loop* of the search process (Nurmela, 1993). For example, if the initial temperature of the algorithm is 1000°, perhaps 300 iterations are conducted using this temperature. When 300 iterations have passed, the temperature may decline to 999° (using a 0.999 cooling rate), where 300 more iterations of the algorithm would be conducted. Nurmela (1993) suggested that the inner loop within a temperature level might be at least as long as the size of the potential neighborhood of moves. In the case of a forest harvest scheduling problem, the inner loop might be (management units × (harvest choices – 1)), where *harvest choices – 1* accounts for the fact that one choice may have already been assigned to a management unit (stand). More than likely, some experimentation with different lengths of the inner loop would be necessary to understand the reasonable range.

3.3 Temperature Value

An important component in a simulated annealing search process is the temperature, as it controls the behavior of the process, transitioning from diversification during the early stages of a search to intensification during the later stages of a search (Franzin & Stützle, 2019). At the initiation of a search process, the temperature is often declared by the user. From the value declared, the temperature decreases as the search process proceeds. The temperature within a simulated annealing search process is really not meant to be degrees (Celsius or Fahrenheit), but rather just a value, yet it is referred to as degrees to remain consistent with the conceptual premise behind the development of this heuristic search process. As was suggested earlier, the temperature is important in determining the probability of acceptance of inferior moves within the search neighborhood, as the temperature is an analogy for the time exerted to solve the problem.

Example

Early in the search process of a maximization problem, an inferior move might be proposed:

$$\text{Proposed solution objective function value} = \$9,000,000$$
$$\text{Current solution objective function value} = \$9,200,000$$
$$\text{Temperature} = 200,000\,^\circ$$

$$\text{Probability} = \text{Exp}(-((9,200,000 - 9,000,000)/200,000)) = 0.367879$$

Therefore, there is about a 36.8% chance that this inferior solution will be accepted, if a random number drawn from a computing device, within the

(continued)

range of 0–1, is smaller than 0.367879. When the temperature cools, the probability of accepting a similar move with the search is much lower.

$$\text{Proposed solution objective function value} = \$9,000,000$$
$$\text{Current solution objective function value} = \$9,200,000$$
$$\text{Temperature} = 100,000\,^{\circ}$$
$$\text{Probability} = \text{Exp}(-((9,200,000 - 9,000,000)/100,000)) = 0.135335$$

Here, the probability of accepting a similar inferior move, where the objective function value will decline by $200,000, is only 13.5%.

The initial temperature for a simulated annealing search process likely will be different for each problem to which the algorithm is applied. It has been suggested that when the initial temperature is too high, a search process may waste computing time by diversifying the search in inefficient areas of the solution space with no incremental improvements on the best solution (Nurmela, 1993). In other words, a temperature level that is too high promotes diversification, which is good in a limited sense, but may cause the search process to wander aimlessly throughout the solution space. An initial temperature that is too low discourages diversification, because the probability of allowing inferior changes to a current solution (moving to a neighboring solution of lower quality) will be very low, and this may result in the generation of lower quality solutions overall because the search process was not allowed to sufficiently explore the various regions of a solution space. The behavior of a simulated annealing search process can be quite different when applied to the exact same problem (Figs. 9.2 and 9.3), depending on how a computer accesses its random number generator or list and the quality of the initial solution.

Although trial-and-error processes have been used to determine a range of initial temperatures that seem to provide better solutions to a problem, it has been suggested that one can approximate the appropriate initial temperature by determining the temperature at which the average change in solutions values allows a certain assumed (at least initially) probability of acceptance of inferior changes to a current solution (Nurmela, 1993). Other approaches have also been used to determine the initial temperature by making it proportional to the maximum difference observed amongst consecutive early solutions (Franzin & Stützle, 2019). Even further, some (e.g., Connolly, 1990) have suggested using a constant temperature that does not cool, or using a temperature scheme that periodically reheats, to encourage further exploration of the solution space.

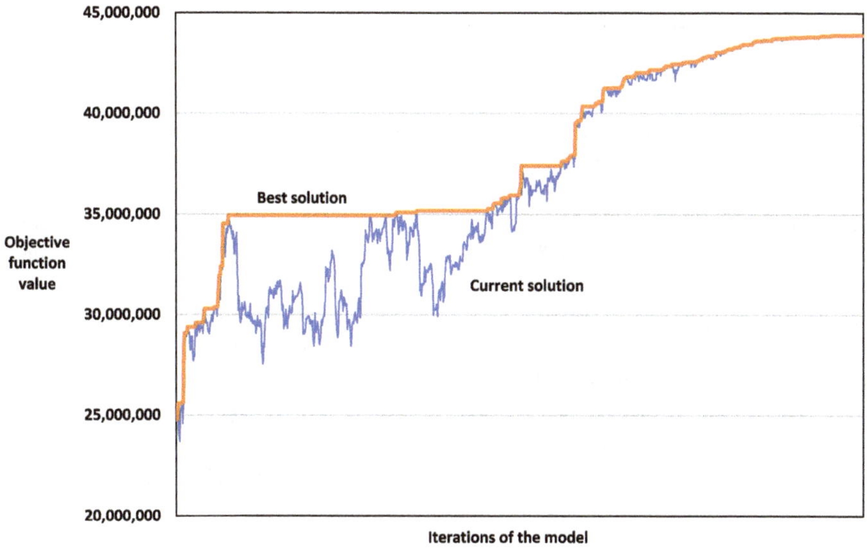

Fig. 9.2 The behavior of a simulated annealing search process when applied to a forest harvest scheduling maximization problem. Here, initial entry into the random number list or generator is based on the computer clock time. Credit: Pete Bettinger

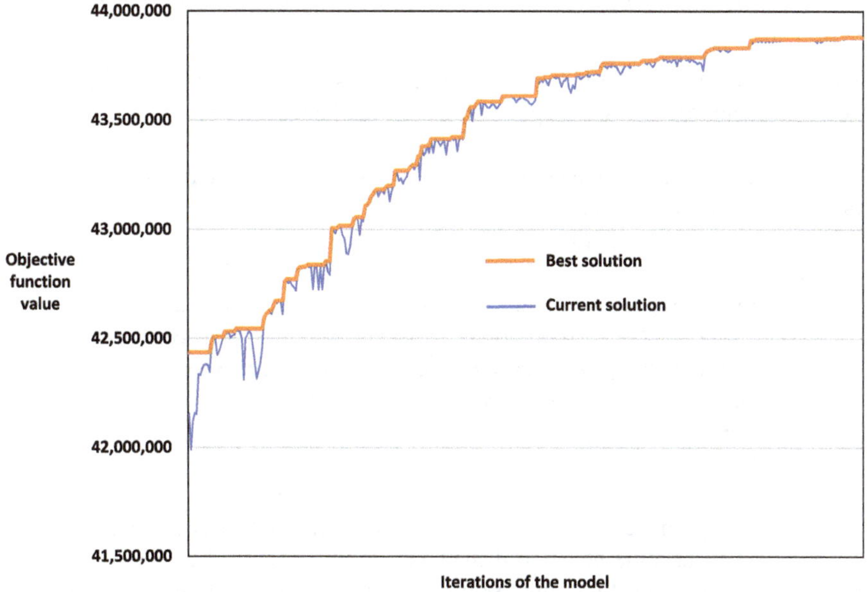

Fig. 9.3 The behavior of a simulated annealing search process when applied to the same forest harvest scheduling maximization problem. Initial entry into the random number list or generator is again based on the computer clock time, yet this run was initiated at a different time than the one in Fig. 9.2. Credit: Pete Bettinger

3.4 Temperature Change

The efficiency of the simulated annealing search process is dependent on careful consideration of the cooling schedule (Hopper & Turton, 2001). An iteration of a simulated annealing search process involves the successful change from one solution S_1 to another S_2 through the random selection of 1-opt, 2-opt or *n*-opt moves, and successful satisfaction of constraints. An unsuccessful iteration either results from constraint violations or from failing the simulated annealing test, which in part depends on the current temperature. The temperature cools off (anneals) as the search process progresses given a few assumptions by the developer of the model. For example, cooling might occur with every attempted iteration of the search process, with every successful iteration of the search process, or with every *m* successful or unsuccessful iterations of the search process. When cooling is expected to occur, a further assumption involves the manner in which the temperature changes. General methods (from Talbi, 2009) for changing the temperature include:

- Reduce it by a constant amount (linear reduction).
- Reduce it by a constant factor (geometric reduction).
- Reduce it by a logarithmic factor (new temperature = old temperature/Log (search iterations)).
- Reduce or increase it by non-monotonic or adaptive factors that depend on search history.

In the example below, every 25 successful iterations of a simulated annealing model the current temperature is multiplied by a constant factor (0.995).

```
   time_to_change_temperature = 25  ' Change temperature every
25 successful iterations
   change_factor = 0.995
   If (proposed solution is feasible) then
     successful_iterations = successful_iterations + 1
     If (successful_iterations = time_to_change_temperature) then
       current_temperature = current_temperature × change_factor
       successful_iterations = 0
     End if
   End if
```

When the change in temperature is (a) a constant amount, it is possible that the temperature can become negative if methods are not employed to curtail the temperature to a zero or positive value. When the change in temperature is (b) a constant factor, the temperature will approach 0 in a non-linear fashion.

Example

As was noted earlier, in cases where the search process wants to move to an inferior solution (from the perspective of the objective function value), there is a higher probability of the proposed solution being accepted when the temperature is relatively high than when the temperature is relatively low (Franzin & Stützle, 2019). As the temperature approaches 0, the probability of accepting an inferior solution during the search process also approaches 0. From the previous example, if the temperature declined to 1000°, the probability of accepting a similar move as before drops to nearly 0%.

$$\text{Proposed solution objective function value} = \$9,000,000$$
$$\text{Current solution objective function value} = \$9,200,000$$
$$\text{Temperature} = 1000\,^{\circ}\text{s}$$
$$\text{Probability} = \text{Exp}(-((9,200,000 - 9,000,000)/1000)) = 1.38\text{E-}87$$

To summarize, as the temperature declines, the probability of allowing large deviations from the current solution (with respect to the objective function value of the current solution) becomes less likely. Therefore, it may be important to monitor the consecutive unsuccessful attempts to change the current solution (as discussed in threshold accepting) and use a mechanism to change the temperature when a large number of these have been observed so that the search process terminates in a timely manner. At least in one case (Nurmela, 1993), a geometric cooling schedule seemed to be the most efficient and robust of the potential cooling processes. Others have suggested that the temperature might be reset back to the initial (or some other state) during the search to promote diversification after prolonged periods of search intensification (Franzin & Stützle, 2019).

3.5 Search Reversion

As with other *s*-metaheuristic search processes, search reversion can potentially improve the quality of results obtained (Bettinger et al., 2015). Search reversion can be viewed as a diversification strategy (Thomas & Salhi, 1998), but it behaves more like an intensification strategy. Ideally, an *s*-metaheuristic search process seeks better quality, feasible solutions in the nearby neighborhood, perhaps along a Markov chain (Frausto-Solis et al., 2021). In search reversion, if after some amount of time the quality of the best feasible solution has not been improved, the search stops, the current solution is replaced by (or reverts back to) the best solution, and the search process begins again from there. A very simple reversion strategy for simulated annealing might involve the following:

```
time_to_revert = 10
Do while temperature > 1
   ...
   revert_count = revert_count + 1
   If (revert_count = time_to_revert) then
     replace the current solution assignments with the best solution
         assignments in memory
     revert_count = 0
   endif
   ...
Loop
```

In the example above, which is similar to the example provided in the threshold accepting chapter, the search process will pause every ten iterations, and the current solution (whatever it looks like) will be replaced with the characteristics or assignments that represent the best solution (whatever that looks like).

As was noted in the threshold accepting chapter, the rate of reversion is a parameter of the search process that needs investigation, as the appropriate reversion rate seems to differ from problem to problem (Bettinger & Zhu, 2006; Bettinger et al., 2015). When compared to search processes that employ no reversion, a search process that employs search reversion usually results in higher quality solutions.

3.6 Termination Rules

The behavior of a simulated annealing search process, as it relates to the change in the temperature value, is important in defining a *stopping rule*. In general, most simulated annealing termination rules involve the temperature declining (the system annealing) to a certain assumed state. Theoretically, the search process should stop when the temperature declines to some minimal value close to zero. For example, if a static, constant change in temperature is employed during the search, a stopping rule might involve terminating the search when the temperature drops below 0°. When employing a rate of change strategy to adjust the temperature, the temperature may never reach 0° exactly, nor drop below 0°. In this case, some minimum temperature would be assumed as the stopping rule.

Alternatively, as was suggested for threshold accepting, a stopping rule might be based on whether a certain number of iterations have passed with no improvement in the objective function value (Nurmela, 1993). For problems where movement throughout a solution space is difficult, where there may be a very high proportion of infeasible or otherwise unsuccessful potential moves compared to feasible potential moves, the difficulty can be measured by counting the number of (a) infeasible or otherwise unsuccessful attempts or (b) consecutive infeasible or otherwise unsuccessful attempts. Some guidelines for ensuring that a threshold accepting search process would terminate were provided in Chap. 8, and they apply here as well. In general, if the search seems to be stalling due to the random selection of moves that are infeasible or otherwise unsuccessful, some action may be necessary to force the temperature to change so that the search process can terminate.

```
Do while temperature > stopping rule
   ...
   If potential move is feasible, count = 0
   If potential move is infeasible or unsuccessful, count = count + 1
   If (count = time to change temperature) then
      change temperature
      count = 0
   endif
Loop
```

4 Contemporary Forest Harvest Scheduling Problem

As a simulated annealing search process may require thousands or millions of iterations prior to termination, only one iteration of the process is described here. Similar to the example provided in Chap. 8, assume that a number of assignments of final harvest time periods to timber stands have been made for the southern Arkansas forest described in Sect. 4 of Chap. 2. Perhaps these assignments are represented by the random search solution described in Table 7.1 from Chap. 7 of this book. In this solution, the objective function value was $4,023,263.55, and for this example assume that it represents the best solution located thus far during the search process. Further assume that the current temperature in a simulated annealing search process is 5000°. Within the search process, stand 17 and final harvest period 8 might both be randomly selected as the potential move to make in the local neighborhood of solutions (from stand 17 and final harvest period 7). The objective function value contribution from scheduling stand 17 for a final harvest during time period 7 was $138,124.96. The objective function value contribution from potentially scheduling stand 17 for a final harvest during time period 8 is $133,573.88. Making this random move through the solution space could result in a decline in objective function value of $4551.08 ($138,124.96 - $133,573.88). The probability of making such a move within the local neighborhood of the solution space is

$$\text{Probability} = \text{Exp}(-((138,124.96 - 133,573.88)/5000)) = 0.402$$

This suggests that this potential change to the harvest schedule has about a 40% chance of being permitted. If a random number drawn from the computing device that is managing the search process is smaller than 0.402, the proposed change to the harvest schedule would be permitted. This assumes, of course, that none of the constraints will be violated when this change to the harvest schedule is made. Had the current temperature been 1000°, this particular potential change in the assignment of a final harvest time period to stand 17 would have had a 1% chance of being permitted.

5 Applications

In the literature, there are several examples of applying simulated annealing search processes to forest harvest scheduling problems. Lockwood and Moore (1993) were among the first to illustrate the utility of the search process in developing a forest harvest schedule with wood flow and harvest adjacency constraints. Öhman and Eriksson (2002) describe methods for integrating linear programming with simulated annealing in forest harvest scheduling problems that included wood flow and harvest adjacency constraints. In this case, the heuristic search process was employed to address the spatial constraints, while the exact method was employed to address the non-spatial constraints, under the premise that a heuristic search process acting alone may be more prone to terminate at local optima.

In an interesting analysis of a 16,027 acre (6486 ha) forest located in the Hunan Province of China, Deng et al. (2022) used simulated annealing to produce solutions to a management problem that involved a 100-year time horizon which was partitioned into 5-year planning periods. The objective was to maximize the net present value of wood products produced as well as the net present value of carbon sequestered. Weights were applied to each to address priority, and integer decision variables were assumed. Constraints included a minimum harvest age, wood flow assumptions (a deviation of ± 20% from one period to the next was allowed), harvest levels that were less than growth of the forest, harvest restrictions on steep slopes, and area constraints to encourage the development of a normal forest. The simulated annealing algorithm utilized an initial temperature of 1,000,000°, a 10° final temperature, and a 0.995 cooling rate. The search process also assumed that 300 iterations per temperature level would be conducted.

With the assistance of a simulated annealing heuristic algorithm, Chen et al. (2022) assessed several alternative forest plans for a 26,454 ha (65,367 acre) forest in northeast China where net present value was maximized over a 50-year time horizon that was subdivided into ten 5-year time periods. The potential management actions included several partial cutting options for natural forests and plantations, and some final harvest options for larch (*Larix olgensis*) and Mongolian pine (*Pinus sylvestris* var. *mongolica*) plantations. Management constraints focused on minimum harvest ages, relatively even flow of wood products (15% deviation allowed from one period to the next), maximum clearcut harvest areas (40 ha), an associated clearcut green-up length of 10 years, and minimum habitat suitability index values for forest conditions that favored the needs of the great spotted woodpecker (*Dendrocopos major*). The simulated annealing algorithm utilized a 10,000° initial temperature, a cooling rate of 0.999, an assumption of 300 iterations per temperature, and a terminating temperature of 1°.

In practice, the Patchworks™ (Spatial Planning Systems, 2023) simulated annealing search process has been used to analyze different forest management scenarios for large Crown leases in Canada where deviations from various goals were minimized and standing inventory levels, among many other resource outcomes, were constrained. For example, in using this model to develop a harvest schedule for the C5 Forest Management Unit in southwestern Alberta, which encompasses about 870,300 acres (about 352,200 ha) of land, several different management alternatives focused on age class design and desired wood flow levels (Government of Alberta, 2010).

6 Summary

Conceptually, the analogy between a solid entity consisting of particles that are cooling (and the temperature observed at each state of the cooling process) and the characteristics of a combinatorial optimization search process may not be easy to understand (Hertz & de Werra, 1987). The nature of this *s*-metaheuristic process, where only one or two changes to a current feasible solution are proposed during each iteration of the search, can promote both diversification and intensification of the search (depending on the circumstances). However, inappropriate search parameters may limit the ability of a search process from escaping a local optimum (Černý, 1985), and thus the ability of the search process to behave efficiently and effectively may be problem-specific. These problems are not unique to simulated annealing, as all *s*-metaheuristic processes can experience these issues when the selected search parameters are inappropriate for the problem. However, what is different about a simulated annealing search process is that the decision rule to accept an inferior, feasible move in the neighborhood is stochastic in nature, whereas a threshold accepting search process and a tabu search process (Chap. 10) generally utilize decision rules that are deterministic in nature.

Take Home Message

Simulated annealing is a search process that enhances a stochastic hill-climbing search process by accepting inferior moves through the solution space while searching for the optimal solution as long the resulting solutions are not too much worse than the previous (or best) solution held in memory. When an inferior move is proposed, a probability of accepting it is determined, and a decision is made to either accept the move or reject the move. Early in the search, when the temperature of the system is high, search diversification is evident. Later in the search, when the temperature is low, search intensification is evident. This ability to accept inferior moves helps the search process avoid becoming trapped at local optima, a common problem with hill-climbing search processes. This ability allows the search process to move to lower quality solutions with the hope of eventually locating other higher quality solutions.

Questions

1. One or more of the authors of the seminal paper which many attribute as the basis for simulated annealing (Metropolis et al., 1953) are famous for other works. Who are they, and what made them noteworthy?
2. Rather than a steady decline in the temperature (the annealing) during a simulated annealing search process, if the temperature were allowed to vacillate (get hotter, get colder, get hotter, etc.) what impact might this have on the behavior of the search process?
3. What process would you employ to change the temperature of a simulated annealing search process as it navigates through the solution space? Why?
4. Imagine that you are having a friendly conversation with a family member who is unfamiliar with operations research. Describe the simulated annealing search process in simple terms, in five short sentences or less.
5. Philosophically, does the randomness assumed in the moves attempted during a simulated annealing search process improve or inhibit the search for the optimal solution?

References

Bettinger, P., Demirci, M., & Boston, K. (2015). Search reversion within *s*-metaheuristics: Impacts illustrated with a forest planning problem. *Silva Fennica, 49*(2), Article ID 1232.

Bettinger, P., & Zhu, J. (2006). A new heuristic for solving spatially constrained forest planning problems based on mitigation of infeasibilities radiating outward from a forced choice. *Silva Fennica, 40*(2), 315–333.

Broda, E., & Gay, L. (1983). *Ludwig Boltzmann. Man, physicist, philosopher.* Ox Bow Press.

Brooks, S. P., & Morgan, B. J. T. (1995). Optimization using simulated annealing. *The Statistician, 44*(2), 241–257.

Černý, V. (1985). Thermodynamical approach to the traveling salesman problem: An efficient simulation algorithm. *Journal of Optimization Theory and Applications, 45*(1), 41–51.

Chen, Y., Dong, L., & Liu, Z. (2022). Integrating habitat quality of the great spotted woodpecker (*Dendrocopos major*) in forest spatial harvest scheduling problems. *Forests, 13*(4), Article 525.

Connolly, D. T. (1990). An improved annealing scheme for the QAP. *European Journal of Operational Research, 46*(1), 93–100.

Deng, W., Xiang, W., Ouyang, S., Hu, Y., Chen, L., Zeng, Y., Deng, X., Zhao, Z., & Forrester, D. I. (2022). Spatially explicit optimization of the forest management tradeoff between timber production and carbon sequestration. *Ecological Indicators, 142*, Article 109193.

Franzin, A., & Stützle, T. (2019). Revisiting simulated annealing: A component-based analysis. *Computers & Operations Research, 104*, 191–206.

Frausto-Solis, J., Hernández-Ramírez, L., Castilla-Valdez, G., González-Barbosa, J. J., & Sánchez-Hernández, J. P. (2021). Chaotic multi-objective simulated annealing and threshold accepting for job shop scheduling problem. *Mathematical and Computational Applications, 26*(1), Article 8.

Glover, F. (1986). Future paths for integer programming and links to artificial intelligence. *Computers & Operations Research, 13*(5), 533–549.

Glover, F., & Greenberg, H. J. (1989). New approaches for heuristic search: A bilateral linkage with artificial intelligence. *European Journal of Operational Research, 39*(2), 119–130.

Government of Alberta. (2010). *C5 forest management plan 2006–2026*. Government of Alberta.

Gubernatis, J. E. (2005). Marshall Rosenbluth and the Metropolis algorithm. *Physics of Plasmas, 12*, Article 057303.

Hertz, A., & de Werra, D. (1987). Using Tabu search techniques for graph coloring. *Computing, 39*(4), 345–351.

Hopper, E., & Turton, B. C. H. (2001). A review of the application of meta-heuristic algorithms to 2D strip packing problems. *Artificial Intelligence Review, 16*(4), 257–300.

Kirkpatrick, S., Gelatt, C. D., & Vecchi, M. P. (1983). Optimization by simulated annealing. *Science, 220*(4598), 671–680.

Lee, I. (2001). Artificial intelligence search methods for multi-machine two-stage scheduling with due date penalty, inventory, and machining costs. *Computers & Operations Research, 28*(9), 835–852.

Lin, C. K. Y., Haley, K. B., & Sparks, C. (1995). A comparative study of both standard and adaptive versions of threshold accepting and simulated annealing algorithms in three scheduling problems. *European Journal of Operational Research, 83*(2), 330–346.

Lockwood, C., & Moore, T. (1993). Harvest scheduling with spatial constraints: A simulated annealing approach. *Canadian Journal of Forest Research, 23*(3), 468–478.

Metropolis, N. (1987). The beginning of the Monte Carlo method. *Los Alamos Science, 15*, 125–130.

Metropolis, N., Rosenbluth, A. W., Rosenbluth, M. N., Teller, A. H., & Teller, E. (1953). Equation of state calculations by fast computing machines. *Journal of Chemical Physics, 21*(6), 1087–1092.

Michalewicz, Z., & Fogel, D. B. (2002). *How to solve it: Modern heuristics*. Springer-Verlag.

Nurmela, K. J. (1993). *Constructing combinatorial designs by local search*. (Series A: Research Reports. No. 27). Helsinki University of Technology, Digital Systems Laboratory.

Öhman, K., & Eriksson, L. O. (2002). Allowing for spatial consideration in long-term forest planning by linking linear programming with simulated annealing. *Forest Ecology and Management, 161*(1–3), 221–230.

Siedentopf, J. (1995). *An efficient scheduling algorithm based on threshold accepting*. (Report No. 16). Institute of Production Management and Industrial Information Management, University of Leipzig.

Spatial Planning Systems. (2023). *Patchworks™*. Spatial Planning Systems.

Talbi, E.-G. (2009). *Metaheuristics from design to implementation*. John Wiley & Sons, Inc.

Thomas, P. R., & Salhi, S. (1998). A tabu search approach for the resource constrained project scheduling problem. *Journal of Heuristics, 4*(2), 123–139.

Uffink, J. (2014). Boltzmann's work in statistical physics. In E. N. Zalta (Ed.), *The Stanford Encyclopedia of Philosophy*. Summer 2022 *Edition*. Retrieved April 23, 2024, from https://plato.stanford.edu/archives/sum2022/entries/statphys-Boltzmann/

Winkler, P., & Fang, K. T. (1997). Application of threshold-accepting to the evaluation of the discrepancy of a set of points. *SIAM Journal on Numerical Analysis, 34*(5), 2028–2042.

Chapter 10
Tabu Search

Learning Objectives
The objectives of this chapter on tabu search are to help the reader

- Understand the history and motivation of the tabu search process
- Understand the general manner in which tabu search operates
- Understand the use and value of short-term memory and how it might help a search process
- Understand the main limitations associated with the use of tabu search

1 Introduction

Tabu search is a point-based *s*-metaheuristic, which follows the development and evolution of a single solution (forest plan) as it converges, hopefully, upon the global optimum solution. Some have described tabu search as a collection of principles and strategies involving the intelligent use of search memory to locate high quality solutions to problems (Thomas & Salhi, 1998). Tabu search is essentially a deterministic search process, an extension or modification of the hill-climbing algorithm (greedy search) (Nurmela, 1993), whereby after examining a set of potential changes to the current solution, the best choice from these is made. Unlike a hill-climbing algorithm, which only allows improvements to the current solution, tabu search is based on the supposition that selecting a poor change to the current solution may be allowed to help the search process avoid becoming trapped in a local optimum (Glover, 1986). Of the heuristic methods described in this book, tabu search is best described as a slow, thoughtful, rational method for navigating through a solution space in search of high quality solutions.

P. Bettinger, *Forest Harvest Scheduling*, Springer Texts in Business and Economics,
https://doi.org/10.1007/978-3-031-89432-9_10

2 Background

Adaptive memory programming was the original name of the search process now known as tabu search (Glover, 2022). Tabu search is based on general aspects of intelligent problem-solving methods (Glover & Greenberg, 1989). Tabu search moves in a stepwise (iterative) fashion toward the optimal solution of a problem using a special feature that is designed to help the process avoid becoming trapped at local optima (Hertz & de Werra, 1987). The broader application of the word *tabu* to the entire search process was perhaps applied by others because the restrictions that prevent the selection of certain moves along the Markov chain, based on their recent inclusion or exclusion from a solution, were described as *tabu restrictions*, which Fred Glover admits was an impulsive assignment of a label to a process based on the title of a song (Glover, 2022). Some of the early works illustrating the use of tabu search in forest harvest scheduling included Murray and Church (1995), Bettinger et al. (1997), and Brumelle et al. (1998).

Around 1990, two important papers describing the tabu search process were published by Fred Glover from work conducted at the University of Colorado (Glover, 1989, 1990). However, tabu search was perhaps first described a few years earlier in a paper that presented an approach for addressing space planning problems within offices (Glover et al., 1985). As a heuristic search process, tabu search acts like a deterministic hill-climbing search process with two exceptions: (1) the search process does not terminate when there are no more improving moves to make within a solution space, and (2) the search process contains one or more memory mechanisms that act to prevent it from returning too soon to recently visited locations within the solution space (Hopper & Turton, 2001). In contrast to the general method described here in this chapter, the first application of tabu search prevented the repeated selection of moves within a solution space (*cycling*) by penalizing choices that were recently made. The penalties applied decreased in value with each subsequent iteration of the search process. In effect, certain assignments to decision variables were included in a *tabu list* and given a *tabu status* that consisted of a penalty (in objective function value units) that was intended to prevent those assignments from being selected once again until the penalty value had become zero. At that point the potential assignments to decision variables were removed from the tabu list and made fully available once again (without potential penalty) to the search process. For these reasons, Glover (1986) suggested that under certain circumstances tabu search might also be called *weak inhibition search*, since the moves that are tabu at any one point in time during a search are simply a portion of the total moves available to the search, and further the tabu moves become available again after a certain amount of work has been conducted during the search.

3 Process

Although there are options for interjecting randomness into a tabu search process, perhaps this method for solving problems is appealing because it makes logical modifications to a current solution in the search for the optimal solution to a problem. However, like all heuristic methods, locating the optimal solution to a problem in a vast solution space is not guaranteed.

3.1 General Description

After the development of an initial feasible solution (Step 1 in Fig. 10.1) or beginning with a null solution, a tabu search process develops a neighborhood of potential moves, or alternative solutions, which can be reached from the current solution through changes to one or more assignments to decision variables. The neighborhood is designed to recognize (a) each decision variable, (b) each potential change in the assignment to a decision variable, and (c) the potential objective function value that may result when the potential change is actually inserted into the current solution. For example, in Fig. 10.2, the current solution to a problem is represented by a vector where each row represents a decision variable (i.e., row 1 of the vector represents stand 1 of the harvest schedule), and each value represents the assignment. In the case of stand 1, the current solution suggests that it will be harvested during time period 2. The neighborhood contains the potential objective function values achievable by changing the assignment for stand 1 to time period 0 (846), time period 1 (895), or time period 3 (922). All potential changes (moves) are assessed in the local neighborhood. One important decision when constructing a tabu search algorithm involves determining whether constraints are also assessed at the time the neighborhood is developed, to ensure the feasibility of the potential

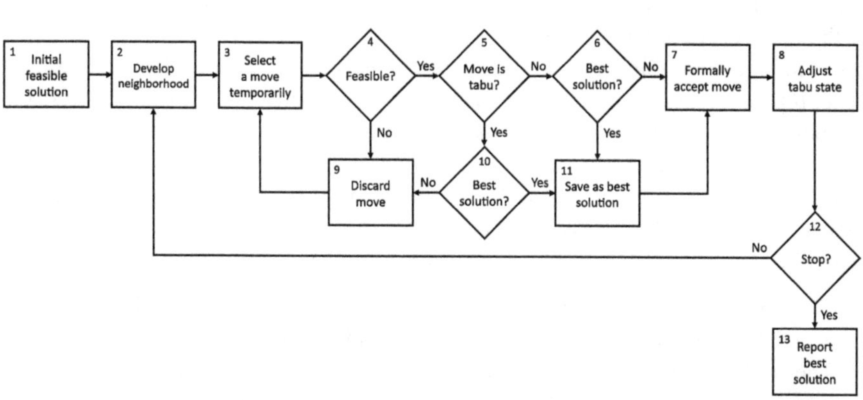

Fig. 10.1 A general tabu search process. Credit: Pete Bettinger

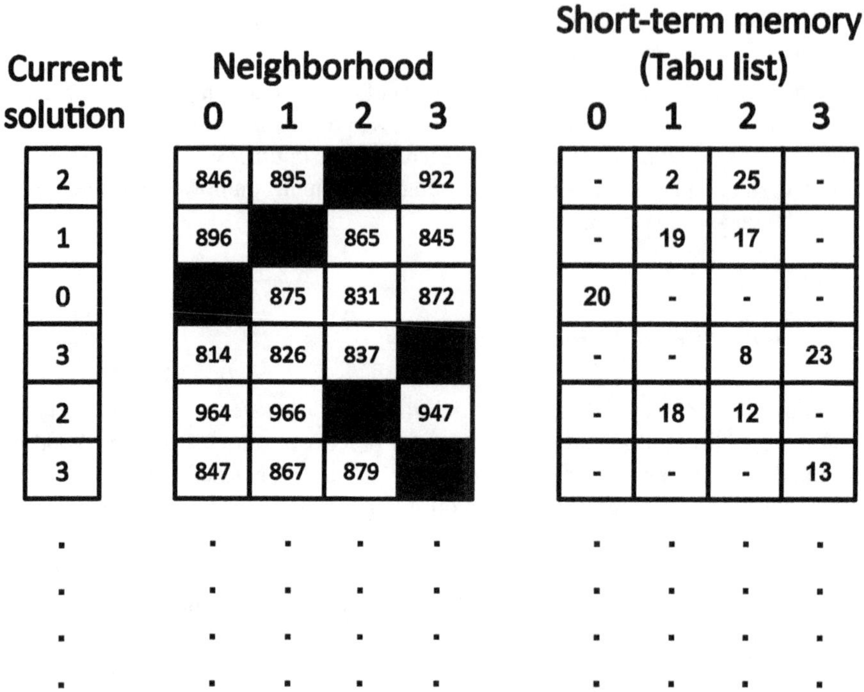

Fig. 10.2 A general structure for a 1-opt tabu search neighborhood and tabu state array. Credit: Pete Bettinger

moves. If an assessment of constraints is time consuming, this effort may not occur until after a move has been selected from the neighborhood (Step 4 in Fig. 10.1).

The best possible move (change in assignment to one or more decision variables in the current solution) is selected from the entire set of potential changes (moves in the neighborhood). Ideally, this potential change leads to a feasible solution, yet if this is not the case, the next best move is selected (and feasibility is again assessed). If the moves selected always lead to higher quality solutions, the search acts like a hill-climbing algorithm. However, tabu search does not terminate when there are no other higher quality solutions which can be reached from a current solution. Rather, the best alternative, even though it may result in a lower quality solution, is selected from the neighborhood. However, if any move selected from the neighborhood has been made recently, it is considered tabu, and the next best move is then considered. Generally, the search process continues in this manner for a pre-defined number of iterations. In sum, a general tabu search process involves these steps:

- Develop an initial feasible integer solution to the problem (Step 1 in Fig. 10.1). This becomes both the current and the best solution to the problem, by default. The initial feasible solution can be developed randomly or deterministically or may simply be represented by a null (empty) set of management actions. For research purposes, a random initial solution is generally preferred.

- Develop the neighborhood of nearby local solutions (Step 2).
- Temporarily select a move from the neighborhood (Step 3).
- Determine whether the temporary proposed solution is feasible with respect to all constraints (Step 4).
  - If the temporary proposed solution is infeasible, discard the move (Step 9) and select another (Step 3).
  - If the temporary proposed solution is feasible, determine whether it is tabu (Step 5).

    If the temporary proposed solution is tabu yet has an objective function value that is higher in quality than the objective function value of the best solution held in memory (Step 10), accept the move and save it as the best solution (Step 11). If not, discard the move (Step 9) and select another (Step 3).

    If the temporary proposed solution is not tabu yet has an objective function value that is higher in quality than the objective function value of the best solution held in memory (Step 6), save it as the best solution (Step 11).

  - If the move has not been discarded, formally accept the move (Step 7).
- Adjust the tabu state of all decision variable/assignment combinations (Step 8).
- Determine whether it is time to terminate the search process (Step 12), and if not, return to Step 2. Otherwise, terminate the search process and report the best solution that was stored in the memory of the computer (Step 13).

3.2 Candidate Moves

A tabu search neighborhood contains the value of neighboring solutions to the current solution. Imagine that a problem contains 1000 timber stands (management units) and recognizes potential management actions across a 20-year time horizon that contains 1-year time periods. If the decision for each stand concerns the year to harvest the timber, then there are potentially 21 choices for each stand (one choice represents not harvesting a stand). However, if we assume that one of these 21 choices is already assigned to each stand, then a 1-opt tabu search neighborhood would involve assessing 1000 × (21–1), or 20,000 potential moves before selecting one. Here, the neighborhood dimension might be

$$\text{number of stands} \times \text{number of potential assignments to stands}$$

As problem size increases, so does the computation time to locate high quality solutions, and since it is often assumed that the entire neighborhood is evaluated, an appropriate data structure is needed to improve computational speed (Thomas & Salhi, 1998). Because of this need to evaluate a neighborhood, tabu search is a

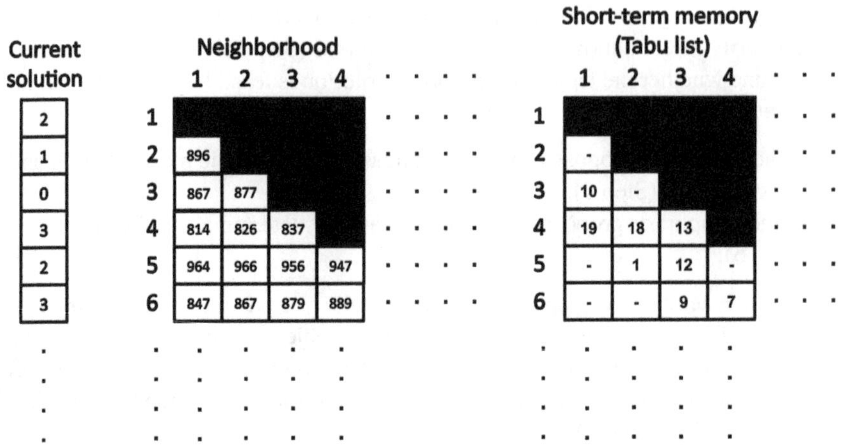

Fig. 10.3 A general structure for a 2-opt tabu search neighborhood and tabu state array. Credit: Pete Bettinger

relatively slow heuristic compared to others, and therefore one opportunity to accelerate a tabu search process may involve restricting the set of potential moves in the neighborhood at each iteration of the model (Thalman et al., 1991). In this respect, it has been suggested that the neighborhood might contain either a random sample of the potential neighbors (Hertz & de Werra, 1987) or that the neighborhood might contain a region-limited set of neighbors (Bettinger et al., 2007).

Employing 2-opt moves within tabu search can require a much more extensive evaluation of a neighborhood if all potential moves away from a current solution are evaluated before one is chosen (Fig. 10.3). In a forest harvest scheduling context, a 2-opt *exchange* move may involve switching (or swapping) the harvest timing (year assigned) of one timber stand with that of another. Here, the neighborhood dimension might be

$$\text{number of stands} \times \text{number of stands}$$

although less than half would need to be assessed.

Example

Imagine beginning with the lowest numbered stand (e.g., stand 1), the harvest period assigned to this stand can be exchanged with the harvest timing of stand 2, or stand 3, or ... stand 1000. Thus, there are 999 potential exchanges involving stand 1. If we then consider stand 2, without conducting a redundant assessment associated with stand 1, the harvest period assigned to stand 2 can be exchanged with the harvest timing of stand 3, or stand 4, or ... stand 1000. Thus, there are 998 potential non-redundant exchanges involving stand 2. If

(continued)

we keep following this logic, there would be 499,500 potential 2-opt moves. In fact, if $n = 1000$ (timber stands) and $r = 2$ (stands involved in the n-opt exchange), we could calculate the combinations of two timber stands whose assignments might be exchanged:

$$\text{Combinations} = \left(\frac{n!}{r!(n-r)!}\right)$$

$$\text{Combinations} = \left(\frac{1000!}{2!(1000-2)!}\right)$$

$$\text{Combinations} = \left(\frac{1000 \times 999 \times 998 \times 997 \times \ldots \times 1}{2 \times 1(998 \times 997 \times \ldots \times 1)}\right)$$

$$\text{Combinations} = \left(\frac{1000 \times 999}{2 \times 1}\right) = 499{,}500$$

In this particular case, there are about 25 times more potential moves to examine in a 2-opt tabu search neighborhood than in the case of a 1-opt neighborhood. Further, once a move has been selected, there are potentially 2 times more constraints to check (because the harvest timing of 2 timber stands will be changed).

As has been noted in research studies, 2-opt *exchange* moves are beneficial for intensifying a search within an *s*-metaheuristic framework (Bettinger et al., 1999; Lee, 2001). However, one interesting aspect of using 2-opt moves is that employing them does not interject any diversity into a schedule of harvests in a forest plan. To be more specific, if there were 47 timber stands scheduled for harvest during year 1 of a tactical plan, there will always be 47 timber stands scheduled for harvest in year 1 in any other version of the plan. The year 1 harvest assignment when using 2-opt exchange moves is simply assigned to another stand, and there is no increase or decrease in the number of timber stands harvested in year 1. It is for this reason that a mixture of 1-opt and 2-opt moves is recommended when conducting a search with a *s*-metaheuristic. Alternatively, one could consider the use of 2-opt *change* moves, which are essentially simultaneous, independent 1-opt moves. This type of process would involve selecting two options from a 1-opt neighborhood, perhaps the best choice and the second-best choice.

3.3 Short-Term Memory Tabu State

The *tabu state* is the amount of time, measured in iterations of a tabu search process, which must pass before a particular choice (assignment to a decision variable) can be selected again from the tabu neighborhood. The tabu state attempts to prevent the search process from selecting moves (changes to the current solution) that would lead the search back to a location in the solution space that had been visited very recently (Hertz & de Werra, 1987). In other words, declaring certain moves as tabu helps prevent a reversal of the current solution to a previously observed solution (Glover, 1990), and thus the search process inhibits certain moves in an adaptive manner using the concept of *short-term strategic forgetting* (Glover & Greenberg, 1989). In Fig. 10.2, the tabu list is the same size (rows, columns) as the neighborhood. Further, the scheduling of a harvest in stand 1 during time period 1 is considered tabu for 2 more iterations of the tabu search process and the scheduling of a harvest in stand 1 during time period 2 is considered tabu for 25 more iterations of the tabu search process.

The tabu state can either be considered a constant number of iterations of a tabu search process or be allowed to vary randomly between 1 and *n* iterations of a tabu search process. Consider this: at some point during a search process, a change was made to the harvest timing of stand 71, moving it from time period 5 to time period 6. This choice (stand 71, period 6) is then considered tabu (forbidden) for *n* iterations of the search process (assume here a constant value of 25). One iteration of the tabu search process later, a change was made to the harvest timing of stand 71, moving it from time period 6 to time period 2. Then, one iteration later, a change was made to the harvest timing of stand 71, moving it from time period 2 to time period 3. Finally, a change was *attempted* to the harvest timing of stand 71, *attempting* to move the harvest from time period 3 back to time period 6. Since only 3 iterations have passed since stand 71 was scheduled for harvest during time period 6, this assignment to stand 71 is considered tabu.

The tabu state of an assignment to a decision variable is applied when the assignment is used successfully to change the current solution to a neighboring solution. In a general implementation of tabu search, the tabu state of an assignment to a decision variable also decreases as each successful iteration of a tabu search process occurs. The tabu state of any particular assignment to a decision variable becomes

$$\text{Tabu state}_{\text{now}} = \text{Tabu state}_{\text{previously}} - 1$$

with each successful iteration of a tabu search process, until Tabu state $_{\text{now}} = 0$. The tabu status of decision variables and assignments is therefore dynamically updated as a search progresses (Thalman et al., 1991). At this point the assignment to a decision variable is not considered tabu anymore. After a feasible neighbor from the tabu neighborhood has been selected, the following type of process might be conducted when using 1-opt moves to adjust the tabu state of all decision variables.

```
If (feasible iteration) then
   For stand = 1 to last stand
      For harvest period = 0 to last potential harvest period
         If tabu_state(stand, harvest period) > 0 then
            tabu_state(stand, harvest period) = tabu_state(stand,
harvest
               period) - 1
         Endif
      Next harvest period
   Next stand
Endif
```

Notice that in this case, where the search process selects a stand and a new harvest period (or to unschedule a harvest event) to modify the current solution, the *harvest period* begins with period 0 (no harvest) in the For-Next loop since it represents a valid option for a stand.

One would have to decide how to handle effectively the tabu state of decision variables when 2-opt moves are employed. One might (a) simply ignore the tabu state, (b) assign a new tabu state to each of the two timber stands involved in the 2-opt move, or (c) assign a new tabu state to the two timber stand/harvest period assignments involved in the 2-opt move. The latter of these options could utilize the same tabu state array structure that is used for the 1-opt move process (number of stands × number of potential assignments to stands). The second of these options might involve a tabu state array structure similar to the 2-opt neighborhood (number of stands × number of stands) if one were utilizing a 2-dimensional array. Alternatively, the second of these options could utilize the same tabu state array structure that is used for the 1-opt move process by assigning a new tabu state to all choices associated with the two timber stands. The first of these options (ignore) may be the easiest to employ, yet it defeats the purpose of tabu search. The tabu state when using 2-opt *change* moves would be similar to that which is used in conjunction with 1-opt moves, only the tabu state of two choices would be adjusted.

When the tabu state is relatively short, tabu search might be considered a *weak inhibition* process, where only a small portion of the potential moves are tabu (Glover, 1986). However, more complex problems may need longer tabu states (Thomas & Salhi, 1998).

3.4 Potential Cycling of Solutions

With respect to data processing, a cycle has been defined as *an occurrence, phenomena, and interval of space or time that recurs regularly and in the same sequence* (Bureau of the Budget, 1962). One of the most frustrating issues related to the use of tabu search and its deterministic nature, is the *cycling* of solutions.

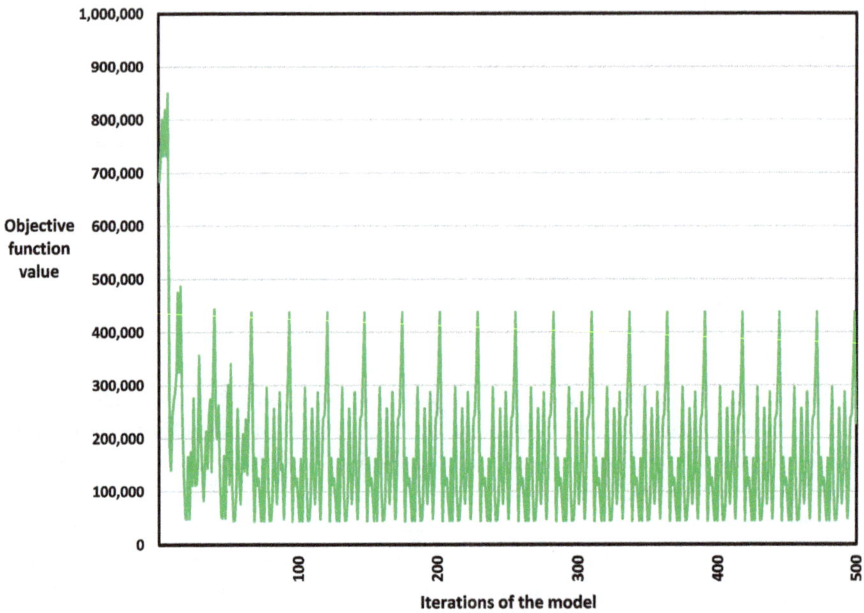

Fig. 10.4 For a minimization problem, an example of the behavior of tabu search when it is cycling amongst the same set of solutions (tabu state = 25 iterations). Credit: Pete Bettinger

During a search through the solution space, the process may revisit the same *X* solutions over and over with no escape (Fig. 10.4).

Example

Imagine that the tabu state of a search is 5 moves. This sequence of moves might occur:

1. The best move is to change the harvest timing of stand 73 to time period 4
   (after which the tabu state of stand 73/assignment 4 is 5 iterations)
2. The best move is then to change the harvest timing of stand 73 to time period 5
   (after which the tabu state of stand 73/assignment 4 is 4 iterations)
3. The best move is then to change the harvest timing of stand 73 to time period 6
   (after which the tabu state of stand 73/assignment 4 is 3 iterations)
4. The best move is then to change the harvest timing of stand 73 to time period 7
   (after which the tabu state of stand 73/assignment 4 is 2 iterations)

(continued)

5. The best move is then to change the harvest timing of stand 73 to time period 8
   (after which the tabu state of stand 73/assignment 4 is 1 iteration)
6. The best move is then to change the harvest timing of stand 73 to time period 9
   (after which the tabu state of stand 73/assignment 4 is 0 iterations, making it available again)

After this sequence of moves in the local neighborhood, the search process then attempts the following moves:

7. The best move is to change the harvest timing of stand 73 to time period 4, since it is available
   (after which the tabu state of stand 73/assignment 4 is 5 iterations)
8. The best move is then to change the harvest timing of stand 73 to time period 5
   (after which the tabu state of stand 73/assignment 4 is 4 iterations)
9. The best move is then to change the harvest timing of stand 73 to time period 6
   (after which the tabu state of stand 73/assignment 4 is 3 iterations)
10. The best move is then to change the harvest timing of stand 73 to time period 7
    (after which the tabu state of stand 73/assignment 4 is 2 iterations)
11. The best move is then to change the harvest timing of stand 73 to time period 8
    (after which the tabu state of stand 73/assignment 4 is 1 iteration)
12. The best move is then to change the harvest timing of stand 73 to time period 9
    (after which the tabu state of stand 73/assignment 4 is 0 iterations, making it available again)

And again, after this sequence of moves, the search process attempts the following moves:

13. The best move is to change the harvest timing of stand 73 to time period 4
    (after which the tabu state of stand 73/assignment 4 is 5 iterations)
14. The best move is then to change the harvest timing of stand 73 to time period 5
    (after which the tabu state of stand 73/assignment 4 is 4 iterations)
15. The best move is then to change the harvest timing of stand 73 to time period 6
    (after which the tabu state of stand 73/assignment 4 is 3 iterations)

(continued)

16. The best move is then to change the harvest timing of stand 73 to time period 7
    (after which the tabu state of stand 73/assignment 4 is 2 iterations)
17. The best move is then to change the harvest timing of stand 73 to time period 8
    (after which the tabu state of stand 73/assignment 4 is 1 iteration)
18. The best move is then to change the harvest timing of stand 73 to time period 9
    (after which the tabu state of stand 73/assignment 4 is 0 iterations)

This sequence of 18 moves occurs over and over again in this fashion because after five moves, an assignment to a decision variable is no longer considered tabu. This *cycling* of solutions will never end, and the search process wastes the remainder of the allotted iterations doing this over and over again.

The length of the cycle is directly related to the length of the tabu state. To counter this behavior, an additional tabu restriction may simply be applied directly to the decision variable (stand) regardless of assignment (harvest period) to prevent the *shuffling* among neighboring solutions that only involve one stand (Glover, 1990). Another option to prevent cycling might be to monitor the current objective function value, to recognize the point when a pattern of moves emerges, and to enact measures to prevent further cycling. In general, a relatively small tabu state leads to the cycling of solutions (Thomas & Salhi, 1998), therefore a third option may be to use a relatively long tabu state (Fig. 10.5). A fourth option may be to vary randomly the tabu state of assignments to decision variables (Fig. 10.6). This latter strategy is not a standard course of action in straightforward tabu search yet seems to be effective in preventing cycling. Any of these options may help prevent cycling from occurring, yet the latter ensures it, and interestingly, interjects stochasticity into the deterministic search process.

3.5 Aspiration Criterion

The word *aspire* means, in one sense, the ability to seek some goal that is of high value. Imagine, after having developed a tabu neighborhood, and after selecting the best move from that neighborhood, that the selected move is tabu (forbidden). The rules of tabu search would dictate that the selection should be ignored, and that the search process should return to the neighborhood and select the next-best move. However, if the objective function value of the selected (yet tabu) move is better than any other feasible solution that had been located to that point in the search process, the aspiration criterion is employed, and the tabu state of the selected move is ignored, allowing the move to be selected from the neighborhood. So, while the

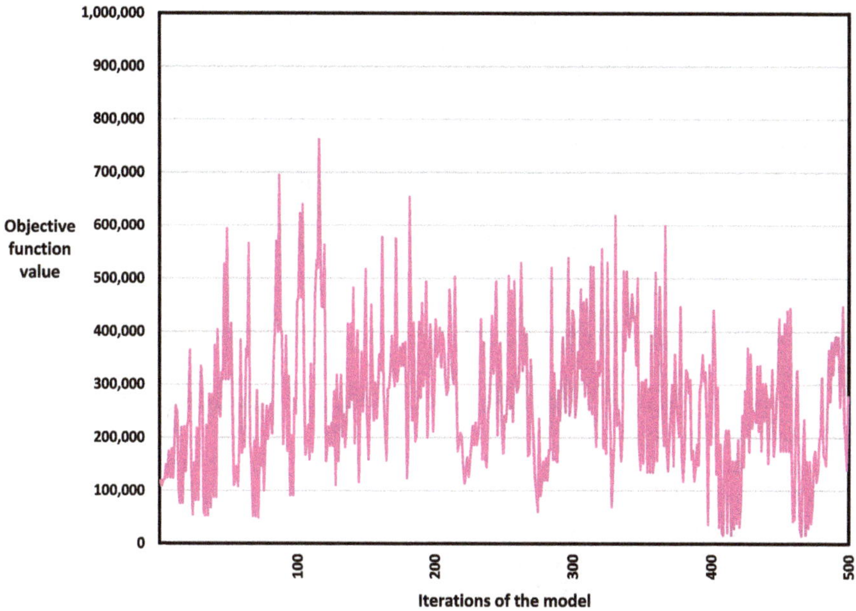

Fig. 10.5 For a minimization problem, an example of the behavior of tabu search when a longer tabu state is assumed (tabu state = 50 iterations). Credit: Pete Bettinger

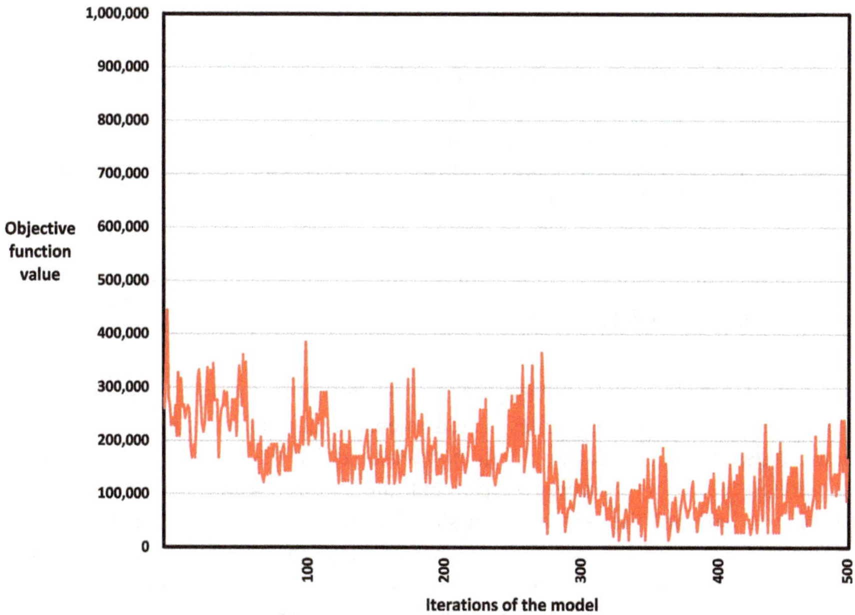

Fig. 10.6 For a minimization problem, an example of the behavior of tabu search when random tabu state is assumed (tabu state = 1 to 25 iterations). Credit: Pete Bettinger

tabu state of a potential assignment might normally suppress the selection of that decision or choice, the aspiration criterion overrides this constraint and allows the move to be selected from the neighborhood (Nurmela, 1993). The search process then replaces the best solution stored in memory with the proposed solution, replaces the current solution with the proposed solution, and updates the tabu state for the move (decision variable and assignment) that was selected from the neighborhood (Steps 10, 11, 7, and 8 in Fig. 10.1). To address each of these actions, a 1-opt tabu search heuristic process might be designed with this logic:

```
    neighboring solution = a feasible move selected from the
neighborhood representing a different assignment to one or more timber
stands
    solution_value = objective function value of the neighboring
solution
    best_solution_value = objective function value of the very best
solution stored in memory
    if (neighboring solution is tabu) then
       If (solution_value > best_solution_value) then
          best_solution_value = solution_value
          current solution = neighboring solution
          best solution = neighboring solution
          tabu state (stand, harvest period) = tabu_state
       Endif
    Endif
```

The use of an aspiration criterion in a tabu search process seems like a logical course of action for taking advantage of the presence of forbidden moves that are of high quality. However, use of an aspiration criterion may not be necessary in order for a tabu search process to work effectively, as some tabu search algorithms lack this mechanism (Goścień et al., 2015).

3.6 Search Reversion

As with threshold accepting and other *s*-metaheuristics, search reversion is an interesting concept that promotes search intensification around high quality solutions. After some computational time (measured by number of search process iterations) the search process stops, and the current solution is replaced by (or reverts back to) the best solution stored in memory. As was noted in earlier chapters, the rate of reversion, or number of iterations that pass before the best solution replaces the current solution, is a parameter of the search that needs investigation. A short reversion rate may not allow a significant amount of diversification to occur during a search, and a long reversion rate may limit intensification around high quality solutions during a search. The appropriate reversion rate seems to differ from problem to problem (Bettinger & Zhu, 2006; Bettinger et al., 2015).

3.7 Long-Term Memory

A long-term memory mechanism might be employed to track the frequency of selection of decision variable/assignment moves during a tabu search process. A structure such as that used for short-term memory might be designed to hold a running sum of the number of occasions that each decision variable/assignment moves have been selected from the neighborhood. At some point, the search process will pause briefly and attempt to insert into the current solution those moves that have been selected least frequently. The intent of this strategy is to interject some diversification into the search and explore unknown (perhaps) areas of the solution space.

4 Contemporary Forest Harvest Scheduling Problem

As with previous chapters, the focus here is on a problem similar to the one described in Sect. 4 of Chap. 2.

As a tabu search process may require thousands or millions of iterations prior to termination, only one iteration of the process is described here. At the beginning of a single tabu search iteration, much like a deterministic hill-climbing search process, the objective function value of all potential changes to the forest harvest schedule would be assessed prior to selecting one. From the current set of potential assignments of final harvest activities to the 31 stands in the southern Arkansas forest, the change in assignment that contributes the most to the objective function value would be selected. If there are no changes in assignments of final harvest time periods that will positively improve the objective function value, the change in assignment that reduces the quality of the objective function value the least would be selected. Given the harvest schedule presented in Table 7.1, if the final harvest assignment of stand 2 were changed from time period 11 to time period 1, the objective function value would increase by $130,756.65 (see Appendix C for data). This change represents the greatest improvement in objective function value of all potential changes to the current solution. Of course, all potential changes in assignments need to have been evaluated prior to selecting this option. If there are no constraint violations as a result of this potential change, and if this particular move in the local neighborhood (selecting time period 1 as the final harvest assignment for stand 2) is not considered tabu, this change is permitted. Once the assignment has been changed, the idea of selecting time period 1 as the point in time in which to conduct a final harvest in stand 2 would be tabu for a certain number of future iterations of the search process.

5 Applications

Richards and Gunn (2000) provided an interesting application of tabu search to a problem that included both forest harvest decisions and road access decisions. The objective was to minimize lost forest productivity through delayed harvests and to minimize road construction costs. The study area was a 11,102 acre (4493 ha) forest containing 1035 stands and 135 road segments. The time horizon was 20 years, which was subdivided into four 5-year time periods. Constraints included limits on final harvest size and limits on deviations from wood flow target levels. A strategic oscillation mechanism was employed to allow the search to cross constraint boundaries (violating constraints) while penalizing the objective function value for doing so. The tabu search process also included a mechanism to recognize cycling behavior and adjust the tabu state value accordingly. Once enough timber stands were scheduled for harvest to meet the harvest targets, a special purpose sub-heuristic was employed to assign values to (i.e., build) the road system.

In Akbulut et al. (2017) a tabu search process was applied to help develop feasible harvest plans for two forests. One 4550 acre (1841 ha) forest contained 87 timber stands. The second 39,087 acre (15,818 ha) forest contained 1109 timber stands. In this implementation of tabu search, the objective was to minimize deviations from a wood flow target over (a) a 30-year time horizon that consisted of six 5-year time periods in the case of the first forest, and (b) a 20-year time horizon that consisted of 20 1-year time periods in the case of the second forest. Constraints included limits on final harvest adjacency and green-up and limits on periodic wood flow deviations. The initial solution for the tabu search process consisted of an optimal solution generated by linear programming, where the assignments to decision variables were subsequently modified (made discrete). In some of the cases examined, the use of the adjusted linear programming solution as a starting point (rather than a randomly generated solution) provided higher quality tabu search decisions.

6 Summary

Tabu search might be considered as one of the more rational decision processes for developing solutions to complex problems. Initially, a tabu search process heads directly for a local optimum solution to a problem (Glover, 1986). Once there, other moves are employed in an attempt to break free of local optima by avoiding selections that have been recently made, in some cases to prevent recent moves from being reversed. The tabu list tracks recently selected moves, promoting flexibility in the search and possibly preventing cycling. These aspects of tabu search, remembering recent moves and preventing cycling, act as a filter to direct the search process to other promising regions of the solution space (Romanycia & Pelletier, 1985).

The overhead required for the logic necessary to develop and manage a neighborhood of potential solutions, and to track the tabu status of potential moves (assignments to decision variables), can be computationally extensive. A *Principle of Rationality* might be appropriately associated with this search method, and as well with the branch and bound method described in Chap. 3.

> *The need to make rational integer decisions when moving throughout a solution space may require substantially greater computational work than similar efforts based on irrational, random decisions.*

There are many ways one can enhance the search behavior of a heuristic search process. For example, in Goścień et al. (2015), the number of consecutive iterations of a tabu search process without leading to an improvement in the objective function value is counted. When a threshold has been reached, a diversification routine is applied to direct the search to a different place in the feasible region of the solution space. This form of search diversification is considered a restart mechanism. As with other potential enhancements to a general tabu search process, the use of a restart mechanism may be advantageous only on a problem-specific basis.

Take Home Message

Tabu search is a logical, deterministic search process that enhances a deterministic hill-climbing search process by remembering recent moves through the solution space while searching for the optimal solution. This memory of recent moves helps the search process avoid becoming trapped at local optima, a common problem with deterministic hill-climbing methods, by allowing the search process to move to lower quality solutions with the hope of eventually locating other higher quality solutions.

Questions

1. Why is tabu search generally much slower in producing a final solution to a problem than threshold accepting?
2. Under what conditions would tabu search produce a solution to a problem much faster than threshold accepting?
3. If a potential move (selection of a decision variable and an assignment of action to that variable) is deemed "tabu," how long should it remain "tabu"?
4. If there were no tabu state, what search process might tabu search most closely resemble?
5. In general, what does the aspiration criterion allow the search process to do when it is enabled? Is there an association between the use of the aspiration criterion and artificial intelligence?
6. Imagine that you were asked to develop an "efficient" forest harvest schedule where the objective function was measured in the net present value of the benefits and costs of activities, and the constraints included limits on the size of final harvests, limits on the spatial distribution of thinning activities across the forest, and requirements for a minimum amount and clustered distribution of raptor habitat conditions. Of the search processes discussed thus far in this book, which would you pursue? Why?

References

Akbulut, R., Bettinger, P., Ucar, Z., Obata, S., Boston, K., & Siry, J. (2017). Spatial forest plan development using heuristic processes seeded with a relaxed linear programming solution. *Forest Science, 63*(5), 518–528.

Bettinger, P., Boston, K., Kim, Y.-H., & Zhu, J. (2007). Landscape-level optimization using tabu search and stand density-related forest management prescriptions. *European Journal of Operational Research, 176*, 1265–1282.

Bettinger, P., Boston, K., & Sessions, J. (1999). Intensifying a heuristic forest harvest scheduling search procedure with 2-opt decision choices. *Canadian Journal of Forest Research, 29*(11), 1784–1792.

Bettinger, P., Demirci, M., & Boston, K. (2015). Search reversion within *s*-metaheuristics: Impacts illustrated with a forest planning problem. *Silva Fennica, 49*(2), article id 1232.

Bettinger, P., Sessions, J., & Boston, K. (1997). Using Tabu search to schedule timber harvests subject to spatial wildlife goals for big game. *Ecological Modelling, 94*(2–3), 111–123.

Bettinger, P., & Zhu, J. (2006). A new heuristic for solving spatially constrained forest planning problems based on mitigation of infeasibilities radiating outward from a forced choice. *Silva Fennica, 40*(2), 315–333.

Brumelle, S., Granot, D., Halme, M., & Vertinsky, I. (1998). A tabu search algorithm for finding good forest harvest schedules satisfying green-up constraints. *European Journal of Operational Research, 106*(2–3), 408–424.

Bureau of the Budget. (1962). *Automatic data processing ... Glossary*. Executive Office of the President, Bureau of the Budget.

Glover, F. (1986). Future paths for integer programming and links to artificial intelligence. *Computers & Operations Research, 13*(5), 533–549.

Glover, F. (1989). Tabu search, part I. *ORSA Journal on Computing, 1*(3), 190–206.

Glover, F. (1990). Tabu search, part II. *ORSA Journal on Computing, 2*(1), 4–32.

Glover, F. W. (2022). Unforeseen consequences of "Tabu" choices–a retrospective. *INFORMS Journal on Computing, 34*(3), 1306–1308.

Glover, F., & Greenberg, H. J. (1989). New approaches for heuristic search: A bilateral linkage with artificial intelligence. *European Journal of Operational Research, 39*(2), 119–130.

Glover, F., McMillan, C., & Novick, B. (1985). Interactive decision software and computer graphics for architectural and space planning. *Annals of Operations Research, 5*(6), 557–573.

Goścień, R., Walkowiak, K., & Klinkowski, M. (2015). Tabu search algorithm for routing, modulation and spectrum allocation in elastic optical network with anycast and unicast traffic. *Computer Networks, 79*, 148–165.

Hertz, A., & de Werra, D. (1987). Using tabu search techniques for graph coloring. *Computing, 39*(4), 345–351.

Hopper, E., & Turton, B. C. H. (2001). A review of the application of meta-heuristic algorithms to 2D strip packing problems. *Artificial Intelligence Review, 16*(4), 257–300.

Lee, I. (2001). Artificial intelligence search methods for multi-machine two-stage scheduling with due date penalty, inventory, and machining costs. *Computers & Operations Research, 28*(9), 835–852.

Murray, A. T., & Church, R. L. (1995). Heuristic solution approaches to operational forest planning problems. *Operations Research Spektrum, 17*(2–3), 193–203.

Nurmela, K. J. (1993). *Constructing combinatorial designs by local search*. (Series A: Research Reports. No. 27). Helsinki University of Technology, Digital Systems Laboratory.

Richards, E. W., & Gunn, E. A. (2000). A model and tabu search method to optimize stand harvest and road construction schedules. *Forest Science, 46*(2), 188–203.

Romanycia, M. H., & Pelletier, F. J. (1985). What is a heuristic? *Computational Intelligence, 1*(1), 47–58.

Thalman, N., Sparn, T., Jaffres, L., Gablehouse, D., Judd, D., & Russell, C. (1991). AI techniques for a space application scheduling problem. In J. L. Rash (Ed.), *1991 Goddard Conference on Space Applications of Artificial Intelligence*. (NASA Conference Publication 3110) (pp. 83–93). National Aeronautics and Space Administration, Office of Management, Scientific and Technical Information Division.

Thomas, P. R., & Salhi, S. (1998). A tabu search approach for the resource constrained project scheduling problem. *Journal of Heuristics, 4*(2), 123–139.

Chapter 11
Genetic Algorithm Search

Learning Objectives
The objectives of this chapter on genetic algorithm search are to help the reader

- Understand the history and motivation of the genetic algorithm search process
- Understand the general manner in which a genetic algorithm search process operates
- Understand the use and value of crossover and mutation and how it might help a search process
- Understand the main limitations associated with the use of a genetic algorithm search process

1 Introduction

A genetic algorithm search process can be described as a population-based *p*-metaheuristic. There exists a population (set) of feasible solutions to a problem, and ideally the individuals in the set are different enough to interject diversity into the search for an optimal solution. Each feasible solution in a population is considered a *chromosome*. If one were to consider the assignments made to timber stands as a 1-column vector, then the analogy is fairly true. In this case, the position of each cell in the vector would represent the stand identification number (stand number), and the value within the cell would represent the assignment to that stand. In the spirit of genetic algorithm terminology, each cell (stand) is considered a *gene*, and the value of the assignment (harvest period, management regime) is considered an *allele*. Of the heuristic methods described in this book, a genetic algorithm represents a semi-rational approach for developing a forest plan through the collaborative effort of many individuals (different forest plans).

P. Bettinger, *Forest Harvest Scheduling*, Springer Texts in Business and Economics,
https://doi.org/10.1007/978-3-031-89432-9_11

2 Background

Genetic algorithms attempt to model strategies of natural adaptation (Bosworth et al., 1972). An example of early work describing mathematical methods for the selection of parent solutions from a population and the recombination of them into child solutions was offered by Fraser (1957). In addition to other work on this subject around this point in time, mutation and crossover mechanisms for solving problems were described by Barricelli (1963). Although the ideas and inspiration that help form the foundation of a genetic algorithm search process are founded in the biological sciences, the development of more formal heuristic approaches to solve complex problems with genetic algorithms seem to have been advanced at the University of Michigan in the 1960s through colleagues and students of John Holland. From this group of scientists, Rosenberg (1967) described the role of crossover as a way to sharing genetic information when solving a problem, and Holland (1970) described *adaptive systems* and the idea that a crossover technique may be applied to two different solutions for a problem to develop new solutions that may be higher in quality. Weinberg (1970) further described the idea of a mutation mechanism for these newly developed solutions to interject into them some additional diversity. Although Bagley (1967) may have been the first to formally use the term *genetic algorithm*, Cavicchio (1970) noted that during these times, genetic algorithms were considered an *evolutionary process* and a set of search procedures called *reproductive plans*, *adaptive plans*, and *reproductive adaptive plans*. Others have indicated that they belong to a set of *evolution algorithms* (Glover & Laguna, 1997). These types of search processes have periodically been described in the forestry literature as novel approaches to address complex, contemporary forest planning problems. In one example, Fotakis et al. (2012) described a genetic algorithm search process that included a spatial operator to adjust assignments when sediment produced from timber harvests became a concern; this operator acted as an additional form of mutation in the formation of child solutions.

3 Process

There are many variations on the concept of a genetic algorithm search process, as there are numerous assumptions regarding (a) the population and how (and how many) parent solutions are selected during each iteration, (b) how crossover is conducted, (c) how mutation is conducted on the child solutions, and (d) how the population evolves after feasible child solutions have been created. Much of the discussion below centers on the basic implementation of genetic algorithms.

3.1 General Description

A general genetic algorithm search process first involves the development of a population of different feasible forest harvest schedules. From this population two or more forest harvest schedules are selected and considered *parent solutions*. These parents are then used to create *child solutions* through the transfer of genetic information amongst the parents. The genetic information could simply be the final harvest time period of a few timber stands yet could be much more extensive and include an entire series of management actions (in sum, *management regimes*) for a few timber stands. Once the child solutions have been developed, they may undergo some random mutations (random changes to final harvest periods or management regimes). These two mechanisms, crossover and mutation, are important unique genetic operators of this heuristic search process (Weinberg, 1970). The child solutions are then assessed for feasibility and for fitness (objective function value). If the child solutions are feasible, they may replace one or more parents in the population. At this point, one iteration of a genetic algorithm search process is said to have been completed. Multiple, perhaps hundreds or thousands, iterations of this kind are conducted during a single run of a genetic algorithm search process. The idea is that the population should evolve, and that the very best parent (with respect to the objective function values of parents) within the population would be close in value to the global optimum solution to the problem. However, as with threshold accepting, simulated annealing, tabu search, hill-climbing, and random search heuristics, optimality is not guaranteed. In sum, a genetic algorithm search process (Fig. 11.1) might follow this sequence of steps:

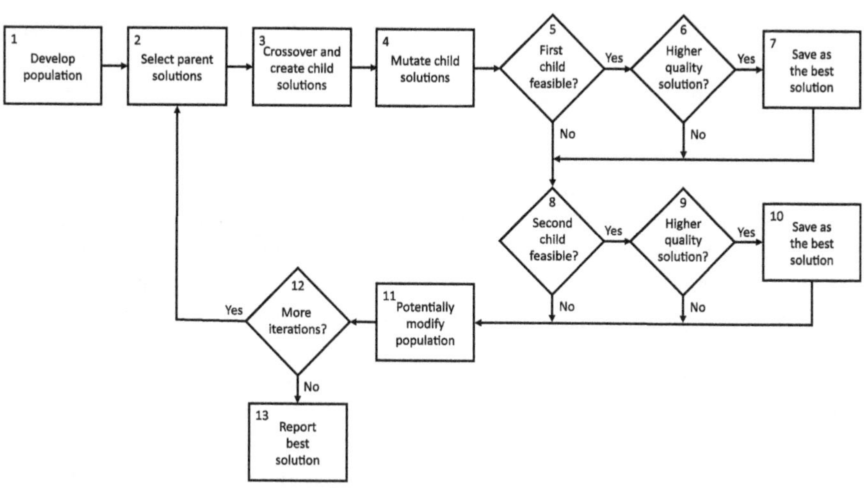

Fig. 11.1 A flow chart representing a general genetic algorithm search process. Credit: Pete Bettinger

- Develop an initial population of feasible integer solutions to the problem (Step 1).
- While the number of iterations < some pre-defined value ...
  - Select parent solutions from the population (Step 2).
  - Determine the crossover point, if one-point or multipoint crossover is assumed, and exchange genetic material to create two child solutions (Step 3).
  - Mutate the child solutions (Step 4).
  - Assess the feasibility of each child solution (Steps 5 and 8).

    If a child solution is not feasible, discard the solution.

    If a child solution is feasible, determine whether it is the best solution to the problem so far (Steps 6, 7, 9, and 10), then insert the feasible child solution(s) into the population (Step 11).

3.2 Population of Parent Solutions

In a *p*-metaheuristic such as a genetic algorithm or particle swarm optimization (Shan et al., 2012), a set of feasible solutions is maintained at each iteration of the search process. The overall quality of the population should be different from one iteration of the search to the next. In this regard, the population evolves as each iteration passes. Ultimately, the one very best, near-optimal feasible solution would be reported by the search process once it terminates.

At the initiation of a genetic algorithm search process, a population of feasible solutions is somehow created. The size of the population is declared by the user of a genetic algorithm, and a feasible solution needs to be generated to represent each member of the population. Imagine, for example, one desires a population size of 500. One strategy might consist of employing a process such as random search to create 500 independent solutions (feasible forest harvest schedules) to represent the initial population. The individuals (the forest harvest schedules) within the population would ideally, as a collection, be diverse in quality. Another strategy might consist of developing the initial population from high quality solutions generated by independent runs of *s*-metaheuristics such as simulated annealing or tabu search. Here, the initial population for the genetic algorithm search process would likely consist of very high quality solutions, which more than likely means the diversity of the initial population may not be very high. A third strategy might consist of systematically collecting the solutions created during a single run of an *s*-metaheuristic. The amount of diversity of the population may depend on the phase of *s*-metaheuristic search (hill climbing, adjustment, fine tuning), but certainly if these solutions were collected consecutively, the initial population for the genetic algorithm search process may not be very diverse.

Ideally, initially, a sufficient amount of diversity is present in the initial population. As each iteration of a genetic algorithm search process passes, one would hope that new high quality (with respect to the objective function value) children would enter the population, and that some poor quality members would leave the population, improving the overall quality of the population. As the search process progresses, one would also hope that the population would evolve to contain many very high quality feasible solutions, yet it would also maintain some diversity. For the search process to be effective, it should avoid two situations: (a) where each member of the population becomes so similar that in effect it is a population of cloned solutions containing very little diversity from one to the next, and (b) where each member of the population is so different that the process does not seem to converge toward near-optimal solutions in the solution space.

The size of the population can remain static throughout the search for an optimal solution, or alternatively the size of the population may change during a search. For certain forestry problems (Lacerda et al., 2023), the population size utilized may have been relatively small (ten members in the case cited) and the stopping criteria relatively short (40 iterations in this case). However, with small populations, important management action assignments may disappear from the solutions that are being developed (Bagley, 1967), and the population may contain insufficient diversity. Alternative designs of the search process could incorporate processes to grow or deplete the size of the population. For example, a search process may begin with 500 members but allow the population of feasible solutions (feasible forest plans) to increase at a certain rate as each iteration passes. In allowing the size of a population to change, one would also need to accommodate a variable amount of computer memory to maintain the increasing or decreasing number of members.

3.3 Selecting Parent Solutions

The motivation of a genetic algorithm is to emulate the reproductive process of biological systems. Therefore, it is common that two parents are selected from the population in order to create child solutions. However, it is not inconceivable that three or more parents can be selected, genetic information exchanged amongst them, and child solutions produced. For example, Weinberg (1970) suggested selecting four parents from a population to create six child solutions. Assuming two parents would be selected from the population, an important consideration in this search process concerns how these two parents are selected. One course of action for selecting both parents is to randomly draw them from the population, where every member has an equal probability of being selected (Fig. 11.2). Another course of action for selecting the parents might be to select the two best parents from the population (Fig. 11.3). Further, one could select the best member from the population as one parent, and a random member as the other (Fig. 11.4). Even further, each member of the population could have an unequal probability of being selected that is based on their fitness (objective function value). An interesting way to select the two

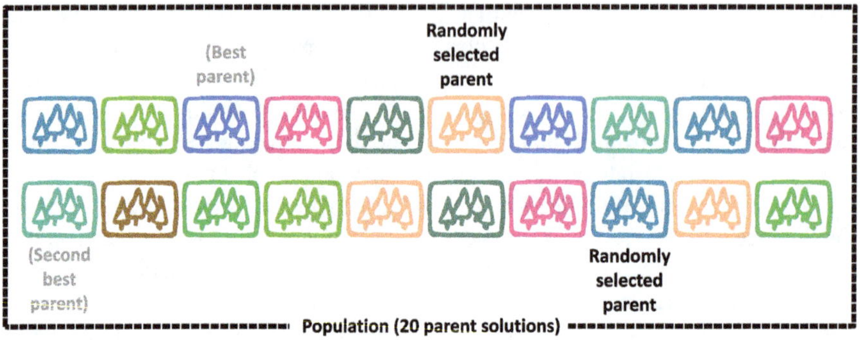

Fig. 11.2 A random selection of parents from the population. Credit: Pete Bettinger

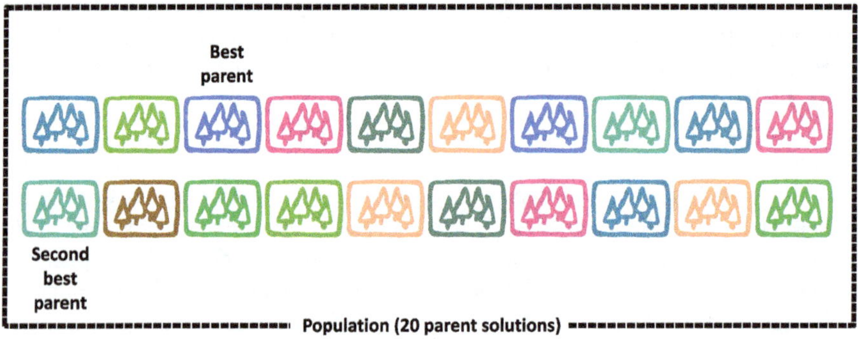

Fig. 11.3 A selection of the two best parents from the population. Credit: Pete Bettinger

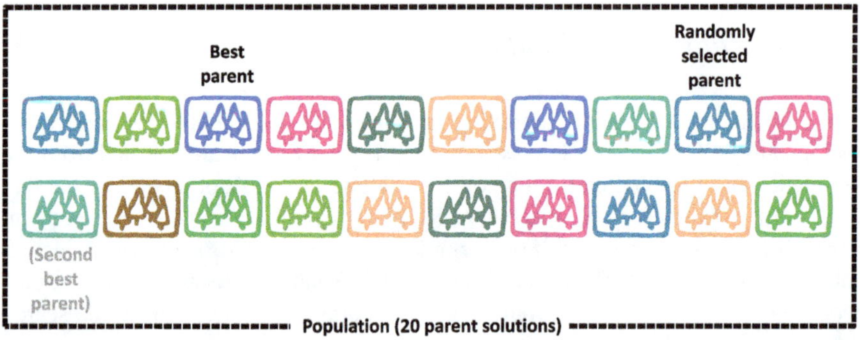

Fig. 11.4 A selection of the best parent and a random parent from the population. Credit: Pete Bettinger

parents is to concoct a tournament, where two subsets, say two 25-member sets, are randomly selected from a population, and the best member (based on fitness) from each subset is put forward as a parent solution. As an example, Ducheyne et al.

(2004) described a forest harvest scheduling problem that used a binary tournament selection process to choose the parent solutions.

From a forestry perspective, once the process for selecting the parents is built into a genetic algorithm, two feasible solutions (forest harvest schedules) are identified as the parents from which the child solutions (two new forest harvest schedules) would be developed. At this point, we know that there are perhaps two different assignments (harvest period, management regime) for each stand in the forest, one suggested by the first parent, and one selected by the second. One of the assignments for each stand of the two feasible forest harvest schedules will be given to the first child solution and the other assignment to the second child solution. In sum,

- Two parents are generally selected from the population.
- The selection can be based on the fitness of each member of the population, or not.
- The two parents selected are feasible solutions.
- The characteristics (assignments made to the decision variables [stands]) of each parent potentially will be passed to a child.

3.4 *Creating Child Solutions*

A crossover process in a genetic algorithm involves the interchange of information between two chromosomes, or parent solutions (Weinberg, 1970). The crossover process involving two parents selected from the population provides each child solution a portion of the assignments (e.g., harvest periods for timber stands, or management regimes for timber stands) from the first parent and a portion of the assignments from the second parent. For example, with respect to a forest harvest schedule, the activities assigned to stands 11–17 of the first parent may be exchanged with the activities assigned to stands 11–17 of the second parent, creating two new child solutions that contain some aspects (genetic material) of the original two parents. One child solution would contain activities assigned to stands 1–10 and 18-*n* of the first parent solution, and activities assigned to stands 11–17 of the second parent solution. The second child solution would contain activities assigned to stands 1–10 and 18-*n* of the second parent solution, and activities assigned to stands 11–17 of the first parent solution.

A crossover process can be designed as a simple *one-point* event or a *multipoint* event. A one-point crossover event is perhaps the simplest method for exchanging information amongst the parent solutions (Foo & Bosworth, 1972).

Example

Imagine that two forest plans (parents) are selected to contribute to a one-point crossover event (Fig. 11.5). For a one-point crossover event, the position for

(continued)

splitting the parents within a vector representing a solution is randomly selected (Fig. 11.6) and then genetic material is swapped to create two new child solutions (Fig. 11.7). The position for splitting the parent solutions has been called the *pivot point* in this search process (Bosworth et al., 1972). A multipoint crossover event has also been called a *double crossover* (Cavicchio, 1972). In a multipoint crossover event, perhaps two points are selected randomly in each solution (Fig. 11.8) and then limited genetic material is swapped to create two new child solutions (Fig. 11.9). During a multipoint crossover, the inner section that would normally be transferred when creating child solutions might instead be inverted within a single parent (e.g., Parent 1 in Fig. 11.8 becoming 23, 12, 5, 17, **23, 24**, 8, 11, 7, 13) as a form of mutation.

Cavicchio (1972) suggested that crossover may only be successful in helping to solve a problem when there is sufficient variability of solutions in a population. We note in this chapter that crossover only occurs once during an iteration of a genetic algorithm, but the search process is not limited in this respect. However, Bagley (1967) suggested that additional crossover events during a single iteration of the search process may not add any value. In some fields, it is understood that a genetic algorithm will not allow the crossover event to occur if infeasible children would be

Parent plan 1

Stand	Prescription assigned
1	23
2	12
3	5
4	17
5	24
6	23
7	8
8	11
9	7
10	13

Parent plan 2

Stand	Prescription assigned
1	4
2	56
3	2
4	11
5	14
6	12
7	23
8	1
9	4
10	7

Fig. 11.5 Assignments of management prescriptions within two parent solutions. Credit: Pete Bettinger

Parent plan 1

Stand	Prescription assigned
1	23
2	12
3	5
4	17
5	24
6	23
7	8
8	11
9	7
10	13

Parent plan 2

Stand	Prescription assigned
1	4
2	56
3	2
4	11
5	14
6	12
7	23
8	1
9	4
10	7

Fig. 11.6 Randomly selected one-point crossover position. Credit: Pete Bettinger

produced (Glover & Laguna, 1997). However, for contemporary forest management problems, this may prove to be computationally inefficient as it may require extensive computation to understand potential outcomes. More than likely, when exchanging assignments of stands between two forest harvest schedules, some constraint violations will occur in association with wood flow or final harvest adjacency concerns. Thus, a multipoint limited crossover event may be more advantageous than a one-point crossover event when a genetic algorithm search process is applied to a contemporary forest planning problem. This limited amount of genetic material exchanged between the two parents will likely result in a lower number of constraint violations and hence a greater number of feasible child solutions.

3.5 Mutating Child Solutions

A mutation process in a genetic algorithm involves changing information contained in a gene (Weinberg, 1970). For the purposes of this book, a mutation may involve potentially changing, with a very low probability, the assignment of harvest period (or management regime) to a single timber stand. The child solutions that are generated after a crossover process are, in this case, subject to uniform random

Child plan 1

Stand	Prescription assigned
1	23
2	12
3	5
4	17
5	24
6	23
7	23
8	1
9	4
10	7

Child plan 2

Stand	Prescription assigned
1	4
2	56
3	2
4	11
5	14
6	12
7	8
8	11
9	7
10	13

Fig. 11.7 Resulting child solutions after the one-point crossover process. Credit: Pete Bettinger

mutation before their quality is evaluated. The purpose of mutating the child solutions is to interject some diversity into the population. However, the drawback is that mutation can result in an infeasible child solution. Nonetheless, there are several ways to apply mutation, but the basic process is to examine each gene (stand) in the chromosome (forest harvest schedule) and potentially change the assignment (harvest period, management regime) to the stand. This has been termed a *point mutation* in a genetic algorithm (Bagley, 1967).

Example

A straightforward application of mutation would be to begin with stand 1 and determine whether the stand should be mutated (have its assignment changed). A random number would be drawn from the computer's random number list, and if the random number is smaller than the assumed mutation rate, the mutation would occur. Otherwise, no mutation would be applied to the stand and stand 2 would then be considered for a mutation, and so on.

```
mutation_rate = 0.01
For stand = 1 to max_stand
draw a random number
```

(continued)

```
If (random number < mutation_rate) then mutate the stand
Next stand
```

In the example above, the assumed mutation rate is 1%. Thus, there is a 1 in 100 chance that the assignment to a stand will be changed. Since computers produce random numbers in the range of 0–1, there is a 1% chance of drawing a random number less than 0.01. If the mutation rate were 0.05, there would be a 5% chance in drawing a random number less than 0.05, and therefore a 1 in 20 chance that the stand's assignment would be changed.

The mutation that occurs to a stand is often simply a random change from the current assignment (e.g., harvest period, management regime) to some other unassigned option available to the stand (e.g., some other harvest period, some other management regime). Further, the probability of selecting one of the unassigned options (option m from the total number of unassigned options M for a stand) is $m/M \times 100$. However, one can imagine that of the M potential unassigned options, some might be given a higher probability (greater odds) of selection than other options.

Parent plan 1

Stand	Prescription assigned
1	23
2	12
3	5
4	17
5	24
6	23
7	8
8	11
9	7
10	13

Parent plan 2

Stand	Prescription assigned
1	4
2	56
3	2
4	11
5	14
6	12
7	23
8	1
9	4
10	7

Fig. 11.8 Randomly selected multipoint crossover positions. Credit: Pete Bettinger

Child plan 1

Stand	Prescription assigned
1	23
2	12
3	5
4	17
5	14
6	12
7	8
8	11
9	7
10	13

Child plan 2

Stand	Prescription assigned
1	4
2	56
3	2
4	11
5	24
6	23
7	23
8	1
9	4
10	7

Fig. 11.9 Resulting child solutions after the multipoint crossover process. Credit: Pete Bettinger

The mutation rate is usually fixed and declared before the genetic algorithm begins its search for the optimal solution. The mutation rate is important, as a low mutation rate, such as the 1% rate noted above, will result in very little modification to each child solution, while a high mutation rate, such as 10% or more, would result in a significant number of modifications to each child solution. A relatively high mutation rate would make a genetic algorithm search process behave much like a random search process, which may not be beneficial in the quest to locate an optimal solution to a problem. Tests seem necessary to determine the appropriate mutation rate for the problem being addressed. Once the child solutions have been subjected to the mutation process, their fitness (objective function value) and their feasibility are assessed. They then potentially could become a member of the population. If so, the mutation process serves to further add diversity to the population.

3.6 *Inserting Child Solutions into the Population*

The children developed through crossover and mutation would ideally, if they are feasible, be incorporated back into the population. If a child solution is infeasible, one simple course of action may be to discard it. Another course of action may be to attempt to repair the child solution and transition it from an infeasible to a feasible

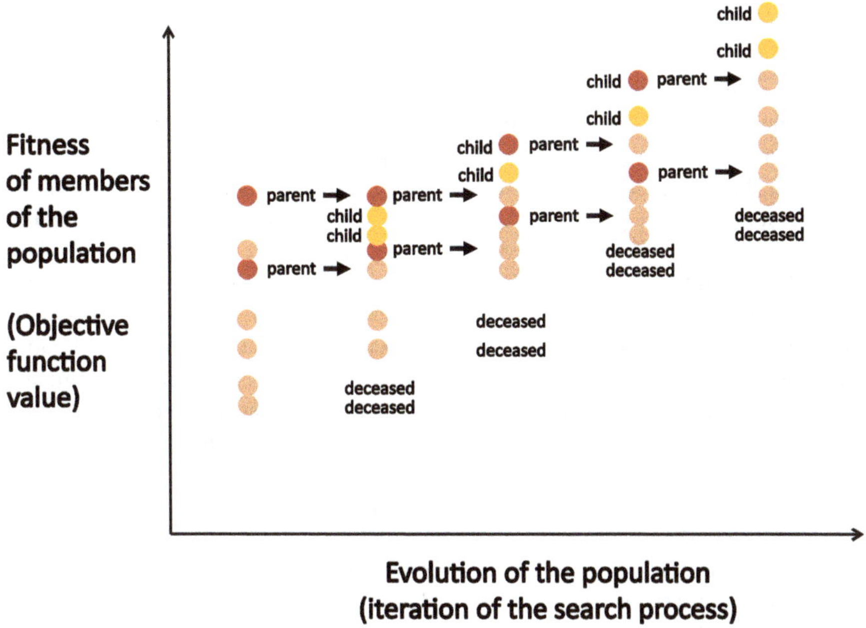

Fig. 11.10 Evolution of the population based on replacing the lowest quality parent solutions with child solutions. Credit: Pete Bettinger

solution. Assuming that the child solutions are feasible with respect to the constraints of a problem, if the fitness (objective function value) of a child solution is higher in quality than the fitness of the lowest quality parent solution in the population, one course of action would be to replace the lowest quality parent with the child solution (Fig. 11.10). Another course of action may be to randomly select parent solutions for removal from the population and replace them with feasible child solutions. The risk in this approach is that very high quality parent solutions may be removed from the population. A third course of action may be to remove from the population the parents that were used to create the child solutions and replace them with the feasible child solutions. Again, the risk in this approach is that very high quality parent solutions (if they were selected to create the child solutions) may be removed from the population. A fourth course of action would be to simply add the child solutions to the population and allow the population to grow.

Beyond the assessment of constraint violations, a genetic algorithm search process may also impose a survivability test upon each child solution to determine whether it would be allowed to join the population. Even when a child solution is deemed feasible, if its fitness (objective function value) does not meet some standard, a rule can be developed to prevent the child solution from entering the population and possibly reducing the overall quality of the population.

3.7 *Termination Rules*

A basic genetic algorithm search process allows a pre-defined number of iterations to occur, after which the best parent located during the search is reported. One iteration of the search process involves the selection of parent solutions, the creating of child solutions, the potential mutation of those child solutions, and the insertion of the child solutions into the population. However, a metric describing the overall quality of the population could be assessed after each iteration, and this information might be used to prompt the termination of the search process. For example, one metric may simply be the average quality (in terms of objective function values) of members of the population. If the search process recognizes that the average quality is no longer improving, perhaps this information can be used as a signal to terminate the search process. As another example, a measure of the variability (again in terms of objective function values) of members of the population could be assessed after each iteration, and this information might be used to prompt the termination of the search process. Measures of variability might include the range, variation, standard deviation, or coefficient of variation of objective function values (some of these are intimately related). If the search process recognizes that the variation has decreased below some arbitrary, acceptable level (indicating low diversity within the population), perhaps this information can be used as a signal to terminate the search process.

4 Contemporary Forest Harvest Scheduling Problem

In accordance with the general problem described in Sect. 4 of Chap. 2, imagine that there exists a population of 100 different forest harvest schedules. Assume here that the genetic algorithm search process is well underway in developing a forest harvest schedule for the southern Arkansas forest. At the onset of one iteration of a general genetic algorithm search process, two parents are selected from the population of forest harvest schedules (Table 11.1). For these two vector representations of harvest schedules, a one-point crossover is applied after the assignment of final harvest time period related to stand 6 and before the final harvest time period related to stand 7, to create two child solutions (Table 11.2). Prior to applying a potential mutation to each of the child solutions, it is apparent that Child 1 results in a slightly higher objective function value (fitness) than either of the parents, and that that Child 2 results in a slightly lower objective function value (fitness) than either of the parents. At this point, with close inspection of the resulting harvest schedules of the two child solutions, it can be determined that Child 1 is infeasible, as stands 5 and 9 are physically adjacent, yet they are now both scheduled for harvest during the same time period (12). Child 2 is feasible in this regard. One by one, each stand in each child solution is then considered for a mutation. If the mutation rate were 1%, a random number would be drawn from the computing device for each of the timber

Table 11.1 Two parent forest harvest schedules selected from a population of schedules during one iteration of a general genetic algorithm search process. Credit: Pete Bettinger

Parent 1			Parent 2		
Timber stand	Final harvest time period	Contribution to objective function value	Timber stand	Final harvest time period	Contribution to objective function value
1	7	275,332.00	1	3	306,872.63
2	11	251,923.13	2	9	273,417.87
3	10	177,914.55	3	2	226,127.19
4	14	152,903.46	4	11	172,854.38
5	2	111,240.43	5	12	80,978.09
6	13	179,587.17	6	7	221,895.90
7	1	88,668.61	7	4	89,157.94
8	4	273,977.60	8	2	285,805.31
9	12	$65,097.06	9	4	64,861.30
10	5	78,820.22	10	9	70,485.81
11	10	139,144.42	11	0	0
12	15	56,461.52	12	15	56,461.52
13	14	62,235.13	13	3	66,670.45
14	9	59,914.47	14	12	57,114.73
15	0	0	15	15	31,805.93
16	2	250,016.62	16	13	163,671.59
17	7	138,124.96	17	14	105,899.58
18	13	75,500.51	18	8	91,284.92
19	10	81,343.44	19	10	81,343.44
20	12	159,198.36	20	5	211,794.80
21	14	177,165.30	21	4	255,747.71
22	9	144,467.52	22	9	144,467.52
23	0	0	23	11	83,052.46
24	6	180,072.47	24	13	140,058.80
25	4	191,773.21	25	15	131,652.22
26	12	27,112.20	26	13	28,032.10
27	1	348,225.12	27	5	306,407.89
28	13	92,784.35	28	8	112,377.49
29	12	95,924.15	29	10	92,519.44
30	12	36,204.00	30	11	35,501.02
31	13	52,131.57	31	14	52,369.37
Fitness		4,023,263.55			4,040,689.40

stands, and if the random number drawn is lower than 0.01, the final harvest time period for the associated timber stand would be randomly changed. From this mutation sub-process, child 1 may become feasible, and child 2 may become infeasible. Whatever the outcome, if either child is feasible after the mutation sub-process, decisions are made to (a) potentially insert either or both child solutions

Table 11.2 Two child forest harvest schedules created during a one-point crossover event during one iteration of a general genetic algorithm search process. Credit: Pete Bettinger

Child 1			Child 2		
Timber stand	Final harvest time period	Contribution to objective function value	Timber stand	Final harvest time period	Contribution to objective function value
1	3	306,872.63	1	7	275,332.00
2	9	273,417.87	2	11	251,923.13
3	2	226,127.19	3	10	177,914.55
4	11	172,854.38	4	14	152,903.46
5	12	80,978.09	5	2	111,240.43
6	7	221,895.90	6	13	179,587.17
7	1	88,668.61	7	4	89,157.94
8	4	273,977.60	8	2	285,805.31
9	12	$65,097.06	9	4	64,861.30
10	5	78,820.22	10	9	70,485.81
11	10	139,144.42	11	0	0
12	15	56,461.52	12	15	56,461.52
13	14	62,235.13	13	3	66,670.45
14	9	59,914.47	14	12	57,114.73
15	0	0	15	15	31,805.93
16	2	250,016.62	16	13	163,671.59
17	7	138,124.96	17	14	105,899.58
18	13	75,500.51	18	8	91,284.92
19	10	81,343.44	19	10	81,343.44
20	12	159,198.36	20	5	211,794.80
21	14	177,165.30	21	4	255,747.71
22	9	144,467.52	22	9	144,467.52
23	0	0	23	11	83,052.46
24	6	180,072.47	24	13	140,058.80
25	4	191,773.21	25	15	131,652.22
26	12	27,112.20	26	13	28,032.10
27	1	348,225.12	27	5	306,407.89
28	13	92,784.35	28	8	112,377.49
29	12	95,924.15	29	10	92,519.44
30	12	36,204.00	30	11	35,501.02
31	13	52,131.57	31	14	52,369.37
Fitness		4,156,508.87			3,907,444.08

into the population of forest harvest schedules, and (b) potentially remove one or more parent solutions from the population of forest harvest schedules. Once these considerations are addressed, one iteration of a general genetic algorithm search has been completed.

5 Application

In a long-term (100-year) forest planning problem involving a 17,381 acre (7034 ha) British Columbia forest, a genetic algorithm was employed to maximize total wood volume scheduled for harvest, subject to minimum harvest age, harvest adjacency and green-up, and wood flow constraints (Liu et al., 2006). The process involved randomly selecting parent solutions from the population, randomly selecting a time period of final harvests within the time horizon, and randomly selecting stands scheduled for harvest in both parent solutions during this time period. The randomly selected stands within one parent solution were exchanged with those randomly selected stands within the other parent. Some repair of the resulting child solutions was necessary as adjacency constraint violations occurred. However, by focusing on one time period, the scheduled (and subsequently exchanged) final harvests satisfied the minimum harvest age constraint, as the harvests would have been feasible with respect to this constraint prior to the exchange. A balancing function was employed within the two child solutions to address the evenness of scheduled harvest levels. This process assigned a final harvest to one or more stands from an unscheduled set of stands to the time period within the time horizon where harvest levels were deemed too uneven. The parent solutions were selected from the population based on a probability that was associated with their fitness (objective function value), which Liu et al. (2006) termed "fitness-proportionate new population selection" to support the notion of *survival of the fittest*. Liu et al. (2006) noted that the inclusion of harvest adjacency constraints within the genetic algorithm required additional computation effort (time) to repair damaged (infeasible) child solutions generated during the during the crossover stage of the process.

6 Summary

Since a genetic algorithm search process attempts to emulate biological reproduction and survival, the mechanisms that it uses to solve complicated problems have attracted the attention of some land management planners and ecologists. The assumptions employed in a genetic algorithm search process are important, as significant crossover and mutation during one iteration of the algorithm can result in large movements through the solution space, while insignificant crossover and mutation can result in small movements through the solution space (Cavicchio, 1972). With respect to contemporary forest harvest scheduling efforts, a carefully constructed genetic algorithm search process may be capable of providing high quality, feasible solutions. Contemporary problems may involve wood flow and harvest adjacency constraints. Certainly, these problems may contain even more complicated wildlife habitat or environmental (soil, water) constraints that employ logic or non-linear relationships and outcomes that are not predictable until they are proposed (given all other actions or assignments that have been scheduled). Careful construction of the search process may require a sophisticated strategy for selecting parents, for conducting multipoint crossover events, for mutating the child solutions, and for evolving the population as a whole. Further, some repair processes for infeasible child solutions may need to be designed to maintain feasibility of the child solutions, and therefore to address the constraints of contemporary forest harvest scheduling problems.

Take Home Message

A genetic algorithm search process is a population-based, stochastic search process that attempts to produce a near-optimal solution to a problem by combining traits of other high quality solutions. Each harvest schedule may contain a set of activities that are indeed the very best (with respect to the objective function and constraints) for the model system under investigation. By creating new harvest schedules from pieces of other high quality harvest schedules, the opportunity to locate a high quality overall harvest schedule seems promising. One common problem with a genetic algorithm search process, as it pertains to a contemporary forest harvest scheduling problem, relates to the infeasibilities that may arise during the crossover and mutation phases. Adjusting the amount of genetic material passed from parent solutions to child solutions, as well as the assumed rate of mutation of the child solutions, may both act to mitigate the amount of constraint violations observed.

Questions

1. In your opinion, what is the best way to select two parent solutions from a population of feasible solutions during the operation of a genetic algorithm search process?
2. What might be some advantages of utilizing a variable-sized population, one that expands and contracts as a genetic algorithm search process progresses?
3. If the child solutions created from the parent solutions during a genetic algorithm search process are both feasible, should they both be incorporated into the evolving population of feasible solutions regardless of their fitness?
4. In your opinion, what might be the best way to implement a mutation process upon the child solutions generated during a genetic algorithm search process?
5. Other than a rule that is based on a fixed number of iterations of a genetic algorithm search process (evolutions of the population), how might you design a genetic algorithm search process to terminate?

References

Bagley, J. D. (1967). *The behavior of adaptive systems which employ genetic and correlation algorithms*. (ORA Project 01252). University of Michigan, College of Literature, Science, and the Arts, Department of Communication Sciences.

Barricelli, N. A. (1963). Numerical testing of evolution theories. Part II. Preliminary tests of performance symbiogenesis and terrestrial life. *Acta Biotheoretica, 16*(3–4), 99–126.

Bosworth, J., Foo, N., & Zeigler, B. P. (1972). *Comparison of genetic algorithms with conjugate gradient methods*. (Contractor Report NASA CR-2093). National Aeronautics and Space Administration.

Cavicchio, D. J. (1972). Reproductive adaptive plans. In *ACM '72: Proceedings of the ACM Annual Conference* (pp. 60–70). Association of Computing Machinery.

Ducheyne, E. I., De Wulf, R. R., & De Baets, B. (2004). Single versus multiple objective genetic algorithms for solving the even-flow forest management problem. *Forest Ecology and Management, 201*(2–3), 259–273.

Foo, N. Y., & Bosworth, J. L. (1972). *Algebraic, geometric, and stochastic aspects of genetic operators*. (Contractor Report NASA CR-2099). National Aeronautics and Space Administration.

Fotakis, D. G., Sidiropoulos, E., Myronidis, D., & Ioannou, K. (2012). Spatial genetic algorithm for multi-objective forest planning. *Forest Policy and Economics, 21*, 12–19.

Fraser, A. S. (1957). Selection of genetic systems by automatic digital computers. I. Introduction. *Australian Journal of Biological Sciences, 10*(4), 484–491.

Glover, F., & Laguna, M. (1997). *Tabu search*. Kluwer Academic Publishers.

Holland, J. H. (1970). Hierarchical descriptions, universal places, and adaptive systems. In A. W. Burks (Ed.), *Essays on cellular automata* (pp. 320–353). University of Illinois Press.

Lacerda, T. H. S., de Jesus França, L. C., Lopez, I. L. E., Lacerda, S. L. S., Figueiredo, E. O., Barbosa, B. H. G., Silva, C. S. J. E., & Gomide, L. R. (2023). Multi-objective forest harvesting under sustainable and economic principles. *Journal of Forestry Research, 34*(5), 1379–1394.

Liu, G., Han, S., Zhao, X., Nelson, J. D., Wang, H., & Wang, W. (2006). Optimisation algorithms for spatially constrained forest planning. *Ecological Modelling, 194*(4), 421–428.

Rosenberg, R. (1967). *Simulation of genetic populations with biochemical properties*. (ORA Project 08333). University of Michigan, College of Literature, Science, and the Arts, Computer and Communications Sciences Department.

Shan, Y., Bettinger, P., Cieszewski, C., & Wang, W. (2012). Pitfalls and potential of particle swarm optimization for contemporary spatial forest planning. *Forest Systems, 21*(3), 468–480.

Weinberg, R. (1970). *Computer simulation of a living cell: Interdisciplinary synergism*. University of Michigan, College of Literature, Science, and the Arts, Computer and Communications Sciences Department.

Chapter 12
Final Thoughts on Forest Harvest Scheduling

Learning Objectives
The objectives of this summary chapter are to help the reader

- Understand why processing efficiency may be important for forest harvest scheduling efforts
- Understand how forest harvest scheduling can be inclusive of other natural resource interests
- Understand how feasibility might be accommodated in a heuristic search process and how this differs from the manner in which exact search processes view constraint violations
- Understand how search process outcomes can be validated
- Understand how the legitimacy of forest harvest scheduling efforts has been recently challenged

1 Introduction

Forest harvest scheduling problems, when expressed as mathematical models, and when addressed using a computer-enabled search process, can provide timely and relevant guidance to foresters and facilitate an assessment of alternative plans of action for decision-makers. The reduction of a real world system to a model forest system may concern decision-makers and field foresters alike; however, the effort required to develop a harvest schedule without a mathematical model and a search process (Fig. 12.1) would likely (a) be difficult for the human mind to process, and (b) result in an inefficient plan of action simply because the human mind is incapable of processing quickly the combinatorially extensive number of alternatives.

Our computational problems today are not new, as for a long time, forest land managers have been concerned with using advanced algorithms to estimate potential timber harvest levels and enhance or maintain other public benefits. For example, in 1973 the Washington Department of Natural Resources highlighted their planning

P. Bettinger, *Forest Harvest Scheduling*, Springer Texts in Business and Economics,
https://doi.org/10.1007/978-3-031-89432-9_12

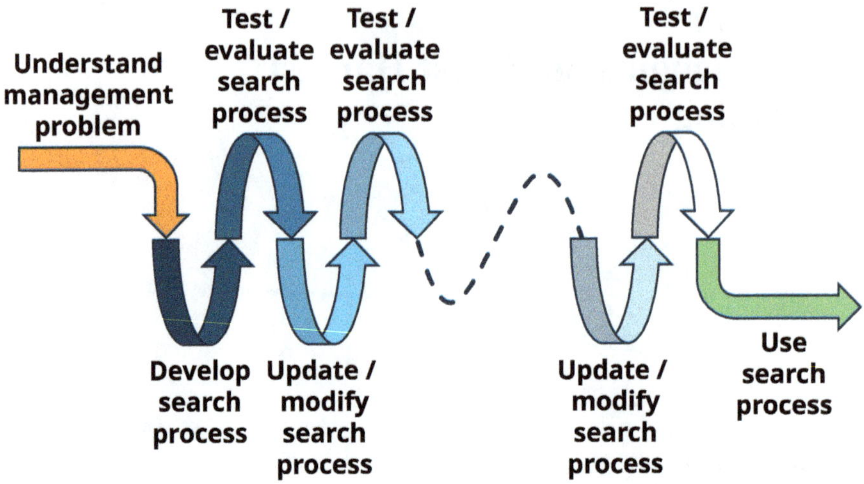

Fig. 12.1 A conceptual model of the developmental and evaluation effort required to develop a search process. Credit: Pete Bettinger

department's operations research activities in their monthly newsletter, explaining simply how economics, forest management, forest mensuration, resource inventories, and data processing technologies are intertwined and capable of assisting in the development of forest management plans that suggest sustainable outcomes will be produced (Washington Department of Natural Resources, 1973). Fifty years later, we still are challenged to provide the guidance foresters seek, and concurrently to provide the assurances of sustainability broader society seeks.

It is interesting that even though there is a lot of hype surrounding artificial intelligence today, as if it were a new concept, many people are unaware that artificial intelligence has been under investigation for over 70 years (Turing, 1950) and has been integral to forest harvest scheduling for decades. Artificial intelligence methods are very useful in many domains of commerce and often they include heuristic search processes (Winograd, 1975). The characteristics of computer models that best relate to artificial intelligence include the ability of the models to both act and to think like a human would in solving a problem. For example, while linear programming and mixed integer programming involve very sophisticated mathematical processes, there arguably is nothing intelligent in these search processes that allow them to function like the human mind, nor to learn from search history or adapt the search behavior to changing conditions (Bettinger et al., 2024). Threshold accepting and simulated annealing, on the other hand, can adapt their search processes based on modifiable acceptance rules, and towards the end of a search they might further emulate human behavior by only looking for the optimal solution in areas of the solution space that show great promise. Tabu search as well could arguably emulate human behavior when employing the memory mechanisms that disallow repeated assessments of options that have been considered recently. And evolutionary computation models may also arguably be considered artificial intelligence through acceptance of non-improving inferior solutions during a search, and subsequent evolution of the population through elimination of inferior solutions.

2 The Quest for Processing Efficiency

There are a number of issues that are central to contemporary forest harvest scheduling efforts; one of these is the quest for processing efficiency. Forest harvest scheduling is conducted to provide guidance to people working in the forest, to help them to understand the appropriate location, timing, and intensity of management actions given the objectives and constraints faced by the landowner. Most of the methods described in this book infer that *appropriate* is synonymous with *efficient*. Efficiency can be described in a financial sense (e.g., try not to waste money) or in a logistical sense (e.g., produce the habitat desired as soon as possible), or both. In addition, a landowner may want to explore quickly many alternative courses of action prior to selecting one. In some cases, this may be accomplished by simply changing the right-hand side of a constraint (i.e., maximum final harvest size, minimum final harvest age, etc.) and comparing outcomes. A well-designed forest harvest scheduling method can assist in many of these areas where processing efficiency (speed of obtaining information) is desired.

3 The Quest for Inclusivity

Although timber production was arguably the main theme of forest harvest scheduling from 1960 to the 1980s, other resource concerns championed by natural resource professionals have been acknowledged in forest harvest scheduling. Therefore, another issue central to contemporary forest harvest scheduling is the inclusivity of natural resource values. With respect to effort exerted in the development of a harvest schedule, any quantitative resource model (even non-linear or logic-based) can be incorporated into a heuristic search process and used as an objective function, constraint, or accounting metric. Exact methods require the use of linear or non-linear equations, and therefore the flexibility afforded by a heuristic search process may provide an advantage here. For direct interaction with an objective function or a constraint, functional relationships between the decision variables and measures of desired outcomes are needed. For example, with respect to a decision variable associated with the final harvest of a stand of trees, a functional relationship with a habitat suitability model might consist of a shared interest in the basal area or canopy cover before and after the management activity was scheduled to occur to that stand of trees.

Example

Examples of incorporating complex functional relationships between management activities and broader resource values into forest harvest scheduling models include the following:

- Hof and Bevers (2000) described Model I harvest scheduling process that used an exact method to minimize sediment produced from management prescriptions applied to timber stands, subject to minimum scheduled timber harvest volume during each time period of the time horizon. A linkage was made between the timing and location of a scheduled management activity and the estimated amount of sediment that would subsequently be routed to a stream network.
- Bettinger and Boston (2008) described a Model I harvest scheduling process that used a heuristic search process to schedule management activities to timber stands and minimize deviations in scheduled timber volume from timber harvest goals, subject to minimum final harvest ages, maximum final harvest sizes, and a weighted average minimum northern spotted owl (*Strix occidentalis caurina*) habitat condition. A linkage was made between the timing and location of a scheduled management activity and the estimated habitat condition at that location (Fig. 12.2) along with other habitat conditions within 2.4 km of that position.
- Dong et al. (2015) described a Model I harvest scheduling process that used a heuristic search process to schedule management activities to timber stands and with the assistance of an additive utility model, simultaneously maximize timber volume, carbon stocks, and the spatial aggregation of management activities through a clustering index, subject to limits on periodic wood flow and maximum final harvest size. A linkage was made between the timing and location of a scheduled management activity, the impact of this activity on carbon sequestered, and the impact of this activity on the desired clumped nature of the management activities.
- Zengin et al. (2015) described a Model I harvest scheduling process that used an exact method to maximize water yield and wood production from management prescriptions applied to timber stands, subject to minimum scheduled timber harvest volume during a thinning activity in a stand and unit restriction adjacency of final harvest activities. A linkage was made between stand condition (which is affected by management activities), ground slope, soil conditions, and water yield potential.

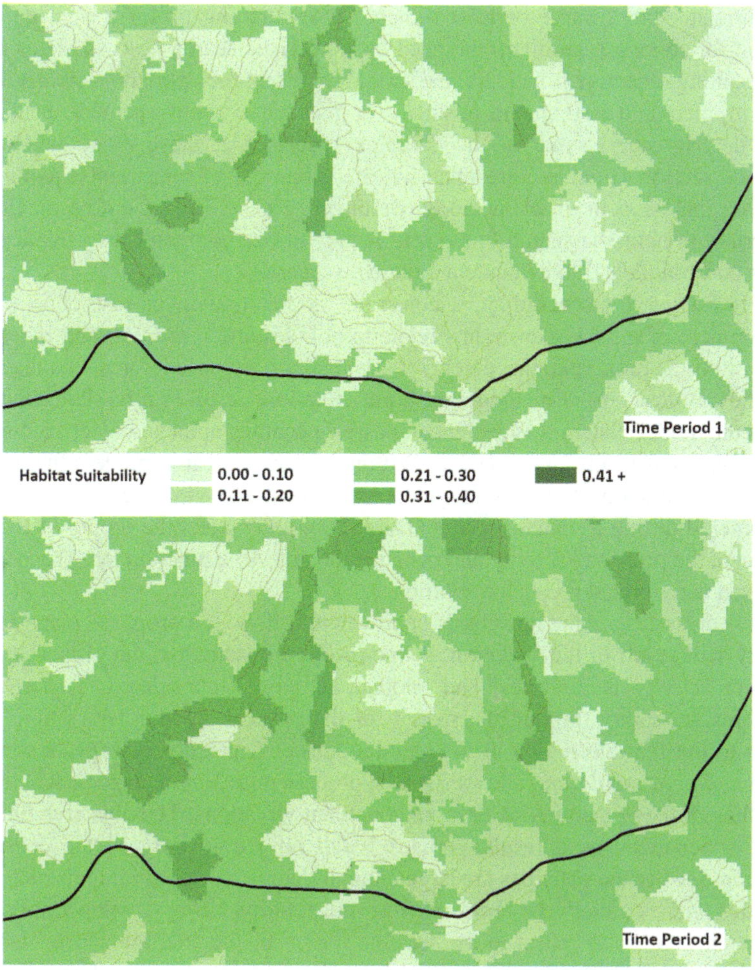

Fig. 12.2 Changes in estimated northern spotted owl (*Strix occidentalis caurina*) habitat suitability between two time periods for a forested area in western Oregon, from the model employed in Bettinger and Boston (2008). Credit: Pete Bettinger

4 The Quest for Feasibility

As the scope of search processes expands, and as the manner in which forest harvest schedules can be developed evolves, the issue of feasibility is central to forest harvest scheduling. There are two domains of feasibility (a) that the management actions suggested by a harvest schedule can be implemented in practice, and (b) that the management actions suggested in the development of a harvest schedule maintain feasibility within the mathematical solution space. This discussion refers to the latter domain. If a forest harvest scheduling problem can be adequately described and formatted as a linear or mixed-integer programming problem (i.e., an exact method for solving a problem) the need for a heuristic search process seems moot. The resulting objective function value should represent the optimal feasible solution to the problem and assignments to stands (or management units) provide the harvest schedule. However, some of the complex models like those mentioned in the previous section could require extensive development of constraint equations to facilitate inclusion in exact methods (Bettinger, 2023). This is not as much of a concern with heuristic search processes, as they utilize computer language logic to assess constraints and to evaluate alternative scenarios.

In forest harvest scheduling, a constraint is a limitation or restriction on the development of a plan. Constraints relate to policies and resources, and inherently to the timing, cost, and the extent of planned or proposed forest management activities. With an exact method, constraint violations result in infeasible solutions, unless the model system was designed to assert penalties upon an objective function when constraint levels are exceeded. Within a heuristic search process there are at least three common strategies for addressing constraint violations once they are encountered during a search process (Talbi, 2009): (1) reject the proposed activity, (2) accept the proposed activity and penalize the objective function, or (3) accept the proposed activity and repair the solution.

A reject strategy is the most basic approach to follow. If a proposed management activity will result in the violation of one or more constraints, the proposed management activity is discarded and another activity is considered. This strategy maintains feasibility through the search for the optimal solution by confining the search to the feasible region of the solution space. A penalty strategy, sometimes called *strategic oscillation*, allows a search process to accept management activities even though they result in one or more constraint violations. The objective function value is penalized as a result, perhaps based on the severity of the constraint violation. Ideally the penalties are large enough that the corresponding management activities will eventually be removed from the solution being developed, allowing the search process to return to the feasible region of the solution space. A repair strategy recognizes the impact (constraint violation) of a proposed management activity, allows the management activity to be scheduled, then enacts various processes to mitigate the difficulties encountered (as in Bettinger & Zhu, 2006). A repair process may use random or deterministic rules to massage the harvest schedule so that feasibility is re-established even though the proposed management activity initially caused problems with respect to one or more constraints. The repairs may involve selecting alternative management activities for land areas (stand, management units) within the spatial vicinity of the proposed management activity or selecting alternative management activities for land areas that are scheduled for management activities in the same general time frame as the proposed management activity. The strategy selected to repair solutions containing constraint violations provides a heuristic search process with a level of flexibility generally unavailable in exact methods for solving problems. However, the computer logic necessary to accommodate a repair process can be extensive.

5 The Quest for Optimality

In addition to the development of a feasible solution, a further issue central to the use of contemporary forest harvest scheduling methods is the quest for the highest possible quality solution. The performance of a heuristic search process, in terms of the quality of the resulting solution(s) generated, might be the most important concern of planners and decision-makers in an evaluation of the quality of outcomes from a harvest scheduling search process. The lowest level of validation is to examine, in the absence of other information, the performance of a heuristic search process when applied to different management scenarios (Bettinger et al., 2009). If a heuristic search process contains a stochastic element, the solutions generated by a heuristic search process will likely be different each time the search process is used (if access to the random number list is based on a computer's clock; otherwise, the final solutions may be the same each time the process is used). The best, average, and worst solutions can be presented in a self-validation effort, as well as measures of variability (standard deviation, standard error, coefficient of variation, etc.) in the solutions produced. With no other benchmark to compare against, these point estimates provide limited information about the quality of a heuristic. However, at a minimum, the range of solution values and the measures of variability can allow one to understand whether solutions are relatively repeatable (small range, low variability) or not (larger range, high variability). Self-validation can also be used to examine the performance of a heuristic search process when applied across different management scenarios. A set of different solutions could be generated by holding the objective function and all constraints constant and applying the scheduling model to different landscapes or properties. Alternatively, a set of different solutions could be generated by holding the landscape or property constant and changing aspects of the objective function or one or more constraints.

The sincerest examination of the quality of results provided by a heuristic search process is to compare the heuristic search outcomes against the outcomes developed from the application of an exact solution technique. This examination requires that both search processes are designed to solve the same problem. If mixed integer programming was employed as the exact method, this examination would also assume that the optimality tolerance was small enough to assure that the optimal solution to a specific forest planning problem would be located.

Comparing the outcomes of one heuristic search process to the outcomes of a second (or third, etc.) heuristic search process is a reasonable course of action to illustrate the relative ability to produce both the highest quality, and the most consistent solutions. Assuming that the testing procedures are fair, one might compare the overall best solution from each heuristic search process, along with the mean (or median) solution values and some measure of variability. Fairness, however, requires the same amount of attention (code development, code verification, etc.) to be paid to the construction of each heuristic search process. For example, spending 2 months developing one heuristic search process and 2 days on another may pose an unfair advantage for the first. In testing multiple heuristic search processes that are applied to the same problem, some consideration also needs to be given to the amount of time and energy each is allowed in searching for the optimal solution. Computer time, model iterations, fitness evaluations (computations of the objective function value) could all be used to control the extent of searches so that each heuristic expends about the same amount of energy. This aspect of fairness could be difficult to control if, for example, one was comparing the outcomes of an *s*-metaheuristic (e.g., simulated annealing) with a *p*-metaheuristic (e.g., genetic algorithm). A further issue of fairness involves the computer system within which the harvest scheduling models operate, particularly if the average time to generate a solution (second, minutes) is one of the metrics used to compare how well each process works. Perhaps the best situation in this respect is to use a single computer; when multiple computers are used, they would ideally be very similar in terms of RAM and CPU speed.

It has been suggested that an estimate of the global optimum solution to a combinatorial optimization problem might be developed from a set of feasible solutions generated by a heuristic search process. To be clear, the set would contain the *best solution* from numerous independent efforts (instances) of applying a heuristic search process to a management problem. Had each instance began its search with a randomly generated initial solution, in theory one could consider each outcome as an independent sample from a larger population. Once this leap of faith is made, one can attempt to find the best Weibull function that would describe the distribution of solutions. A three parameter Weibull function is often employed for this purpose, as the place where the Weibull curve intersects the *x*-axis is known as the *location parameter*, or *extreme value*. In some circumstances, the extreme value can be used as a proxy for, or *estimate* of, the global optimum solution to a forest planning (or other) problem (Golden & Alt, 1979; Los & Lardinois, 1982). However, independent samples from a heuristic search process that develops poor quality solutions to problems may also contribute to a poor estimate of the global optimum solution to a problem (Boston & Bettinger, 1999).

6 The Quest for Legitimacy

The value of a forest harvest scheduling method to a forest management organization or private landowner has been well established, even though credibility and criticism continue (Janová et al., 2024). The selection of harvest scheduling methods and the outcomes produced from these methods have been challenged in courts of law and therefore the quest for legitimacy of these efforts is important. Although it is not the intent of this section of this chapter to overlay a blanket of pessimism on the forest harvest scheduling process, it seems no matter how much effort has been applied to forest harvest scheduling (or more broadly, forest planning) some element of society will take issue with the approach or the projected outcomes. In some circumstances, the development of a forest harvest schedule or an estimate of sustainable harvest levels is required to adhere to the intent of laws (e.g., State of Montana, 2023) or to comply with forest certification standards. In some social systems where it is permissible, the legitimacy of management plans based on harvest schedules has been challenged within legal systems. Particularly with respect to national (federal) lands in the United States, these challenges often relate to concerns regarding the extent of planned management activities and the associated impact on non-timber forest resources (i.e., planned harvest levels are too high, and wildlife habitat will suffer). However, in the following two legal cases, the opposite was argued, that the land managers (state organizations in these cases) were not harvesting enough timber from state forests based on the forest harvest schedules that they had developed.

6.1 Linn County Vs. State of Oregon

In the United States, the State of Oregon (through the Oregon Department of Forestry) is required to maximize "greatest permanent value" on lands transferred to the state from individual counties. In the early- to mid-twentieth century, some counties transferred ownership of their forest lands to the state in consideration of a percentage of the future timber harvest revenue derived from the forest lands. About 36% of the gross revenue from timber sales on state forest lands is provided to the county where those forests reside, thus this source of revenue is important to certain counties. Around the year 2000, the Oregon Department of Forestry endeavored to develop a long-range harvest scheduling model that required detailed information on the location and status of resources and associated management activities so that potential impacts on important natural resources could be acknowledged and measured. The outcomes from the forest harvest scheduling effort included (a) a recognition of northern spotted owl and marbled murrelet (*Brachyramphus marmoratus*) habitat, (b) increased riparian zone restrictions, (c) maintenance of certain older forest structural conditions, and (d) an estimate of potential timber harvest levels.

In 2016, a lawsuit (Fig. 12.3) was brought forward by several Oregon counties, alleging a *breach of contract* had occurred. The plaintiffs in this case argued that the State of Oregon was not harvesting enough timber from state forests in western Oregon, and therefore all counties that contained state forests were due some money based on the perceived *promise* that was made around 2000 regarding potential harvest levels. One main issue noted in the lawsuit was whether the State had properly estimated the amount of sustainable timber harvest with the harvest scheduling model it employed. The plaintiffs asked through the lawsuit's *prayer for relief* for US$1.4 billion. The amount of money requested represented the estimated amount of revenue lost by not harvesting enough timber since around 2000, interest related to this loss, and projected future losses of revenues. Interestingly, the plaintiffs were not asking the State of Oregon to increase the timber harvest amount, they simply wanted the money. Further, if the plaintiffs were successful not only would the lawyers for the plaintiffs receive a share, but the money provided to the

IN THE CIRCUIT COURT FOR THE STATE OF OREGON

FOR THE COUNTY OF LINN

COUNTY OF LINN, on behalf of itself and others similarly situated, Plaintiff, v. STATE OF OREGON; and STATE FORESTRY DEPARTMENT, an Oregon administrative agency, Defendants.	Case No. 16CV07708 **COMPLAINT – CLASS ACTION** **Breach Of Contract** **Claim Not Subject to Mandatory Arbitration** **Statute Setting Filing Fee: ORS 21.160 (Amount in Controversy exceeds $1.4 Billion)** **Jury Trial Demanded**

Fig. 12.3 A portion of the front page of the complaint associated with *Linn County v State of Oregon*. Credit: Pete Bettinger

counties where state forests reside would have to come from citizens in those same, and other, counties of the state.

Of the 36 counties in Oregon, 13 in the western part of the state contain state forest lands. Of these counties, Linn County contains a relatively small amount of state forests as compared to other counties (e.g., Tillamook County). As this case was to be heard by a judge, and the outcome was to be decided by a jury, perhaps the plaintiffs considered the demographics of Linn County were important when deciding where (in which county) to lodge the complaint. Of the people aged 25 and older in Linn County around the time of the trial, about 91% held a high school diploma, about 20% held a bachelor's degree, the median per-capita income was about $32,500 (US Department of Commerce, 2024a), and political support generally favored conservatives. Through the pre-trial depositions and questions posed to expert witnesses during the trial by lawyers associated with both sides of the case (plaintiffs and State), the jury was educated on forest products, forest growth, forest planning, harvest scheduling, and forest sustainability.

The harvest scheduling model was designed to assign management activities to about 23,000 timber stands over a 200 year time horizon that was divided into 20 10-year time periods. Each timber stand was associated with a tree list (current and projected forest conditions). A goal programming approach was developed to maximize net present value while minimizing deviations from timber supply and forest structure goals. Constraints included final harvest adjacency, aquatic and riparian strategies, operability reductions, minimum final harvest ages, and many others. The harvest scheduling model included specific prices for different timber products and tree species, reductions in value due to breakage during logging, regeneration options that recognized a few mature, live trees would remain within final harvest areas, and a large number of wildlife and aquatic system concerns. The management detail included in the model makes it perhaps one of the most complex ever devised. Simulated annealing was used as the search process, as an exact method would not have been possible to devise given the non-linear relationships that were recognized. Ultimately, the weakest link in this harvest scheduling effort, according to a technical review conducted in 2006 (Marmorek et al., 2006), was the forest inventory, as only about one-third of the timber stands had recently been assessed at the time of the harvest scheduling effort. The characteristics of the other two-thirds of the timber stands were imputed forest characteristics derived from the measurements available. This forest inventory situation was not unlike that faced by other large state agencies in the western United States (Mason & Girard, 2013; Stander et al., 2018). Further, even though the harvest scheduling model was perhaps one of the most complex ever devised, a review by state forest managers suggested that some site-specific operational considerations were not recognized, which (had they been modeled) would have reduced the estimated sustainable harvest levels (Lecture, 2002).

In 2019, at the County Courthouse in Albany, Oregon, the State of Oregon lost the case. A jury of 12 Linn County citizens found in favor of the plaintiffs. The State of Oregon appealed the decision to the Oregon Court of Appeals, where in 2022 the decision was overturned (suggesting the State of Oregon did not breach their contract with the 13 counties). The plaintiffs then appealed this decision to the Oregon Supreme Court, which in September 2022 decided that it would not hear the case, affirming the decision by the Oregon Court of Appeals. At the time of the case, the simulated annealing model was state-of-the-art for combinatorial problems. The use of a goal programming approach added legitimacy to the effort by including major goals in the objective function. One might reasonably conclude that (a) given the data available in 2000, and (b) given the manner in which the actual forest management problem was described in the planning model, the scheduled harvest levels should have been viewed as upper bounds on potential real-world harvest levels, to be modified downward given influences on potential management activities arising from local landscape conditions.

6.2 *Concrete School District and Others Vs. State of Washington*

In 1957, the Washington Department of Natural Resources was created by combining nine different agencies and segments of the state government to manage state-owned natural resources in a manner that contributes positively to the quality of life for citizens of the state (Cole, 1971). Forest lands managed by the state are a combination of lands provided by the US government when the state was admitted to the union in 1889, lands that were deeded (transferred) to the state from various counties who acquired them through property tax foreclosures in the mid-twentieth century, and lands gifted to the state. These forest lands are considered held in *trust* by the state, and they are managed to produce revenue for local communities, tax districts, school districts, and others.

This lawsuit was brought forward by several taxing districts in western Washington. These taxing districts receive a portion of the revenues from timber harvests conducted in state-owned forests in their districts, and they alleged a *breach of fiduciary duty* by the state as a greater (or steady) amount of revenues were expected to be received on an annual basis for the management of schools and roads and other local needs. At the time of the case, the demographics of Skagit County (US Department of Commerce, 2024b), where the Concrete School District resides, were similar to Linn County, Oregon, although political support generally favored liberal candidates in the most recent national elections. The plaintiffs (the taxing districts) argued that certain resources (a high quality GIS database of timber stands, a standard growth and yield model, and an industry standard harvest scheduling model) were necessary to assist land managers in developing an implementable management plan that would produce a high amount of revenue on a sustained basis. The plaintiffs further argued that major landowners, such as the Washington Department of Natural Resources, need a reliable, current, and relatively accurate forest inventory to *prudently* plan for and manage forests held in trust for the people of the state. The plaintiffs were asking for relief because they felt that the State of Washington (due to choices made in the development and management of these resources) was unable to achieve the higher, sustainable revenues expected from the 1+ million acres of operable forest it manages on behalf of the people of the state. In the *prayer for relief* of the lawsuit, the plaintiffs were asking for either (a) the removal of the Washington Department of Natural Resources as the manager of these lands, or (b) the return of the ownership of the lands to the counties in which they reside.

The modeling effort conducted in 2019 utilized a strata-based approach (aggregated stand data) as the forest inventory was challenged similarly to the Linn County case. The plaintiffs argued that this was not proper, and that stand-level optimization (along with Model I harvest scheduling) of forest management was the industry standard and the prudent course of action for managing these lands. Along with arguments concerning the management of the GIS databases and decisions to deduct from the operable land base relatively large areas of land deemed critical for habitat purposes, the plaintiffs accused the State of Washington of failing to meet a standard of care for the forest lands held in trust for the people of the state. However, the principle of scale suggests that stand-level optimal solutions will not add up to a forest-level optimal solution due to the forest-level constraints that are not recognized in stand-level decisions. Furthermore, the plaintiffs acknowledged that a commercial harvest scheduling software product utilized by the state was the industry standard, yet this software inherently solves a forest-level problem in Model II fashion (stands are aggregated into strata at the beginning of the time horizon and during the time horizon when certain common activities are scheduled).

In September 2023, shortly before the case was to be heard by a judge and a jury, the plaintiffs were granted a motion to voluntarily end the legal action *without prejudice*. While the case was dismissed by the court, the plaintiffs have the ability to refile it in the future, should they decide to do so. As with *Linn County v State of Oregon*, *Concrete School District and Others v State of Washington* challenged the legitimacy of a state forest management agency to operate as a forest manager, with a special focus on the harvest scheduling processes employed.

7 Summary

A forest plan may be based on the optimal solution from a harvest scheduling search process that was applied to a model of a real-world system, however coarsely or finely defined. A forest plan may also be based on a sub-optimal solution from a harvest scheduling search process if optimality cannot be guaranteed. With respect to forest harvest scheduling, very few aspects of the model system will perfectly describe the real world. The model is an abstract description of human interaction with nature. As with most models of real-world systems, some aspects of the real world either cannot be mathematically described or require too much effort to do so. Once formulated, options for the model system can certainly be explored. These outcomes ideally provide guidance for the people charged with managing the real forest (where to go, what to do, when to do it). However, the real world may present forest managers with unexpected challenges (fires, storms, market fluctuations) that the model system was not designed to recognize, and therefore the plan of action provided to foresters may not be followed exactly as the harvest schedule suggested. Further, issues of efficiency, feasibility, inclusivity, optimality, and legitimacy can challenge forest harvest scheduling efforts. Although forest planning, and specifically harvest scheduling, may be a necessary exercise to conduct, landowners, practitioners, lawyers, and other interested parties in society should keep these issues in mind.

Take Home Message

As a field of work, forest harvest scheduling is challenged by issues concerning processing efficiency, solution feasibility and optimality, resource inclusivity, and developmental legitimacy. There are many reasons why forest harvest scheduling is necessary, from adherence to laws to concerns about financial efficiency and sustainability. Attention paid to the quality of data, the functional relationships between management activities and desired outcomes, and the reasoning behind the selection of search process(es) employed may act to mitigate the concerns others may have about the outcomes from forest harvest scheduling.

Questions

1. Given where we are today with forest harvest scheduling technology, with attention paid to the pace of the developments (computing systems, software, data) that have occurred over the past 60 or 70 years, where will forest harvest scheduling technology be in 30 or 40 more years? What will we be able to do more effectively or more quickly than what we can do today?

2. Do you think that before the end of your career science will have evolved, and functional relationships will have been developed, to more readily model ecosystem services (water, fungi, oxygen, etc.) produced by forests in a forest harvest scheduling effort? Why?
3. Given the behavior of various search processes described in this book, would you be comfortable with a search process strategy that allows navigation through the infeasible region of the solution space? Why or why not?
4. For complex, contemporary forest harvest scheduling problems, would you be comfortable using a heuristic search process to provide a solution? Why or why not?
5. If there was a risk that your forest harvest scheduling effort might be used in a legal proceeding, what might you do to protect yourself from being professionally discredited?

References

Bettinger, P. (2023). Modelling spatial connectivity of forest harvest areas: Exact and heuristic approaches. In C. Grueau, A. Rodrigues, & L. Ragia (Eds.), *Proceedings of the 9th International Conference on Geographical Information Systems Theory, Applications and Management (GISTAM 2023)* (pp. 136–143). SCITEPRESS—Science and Technology Publications.

Bettinger, P., & Boston, K. (2008). Habitat and commodity production trade-offs in Coastal Oregon. *Socio-Economic Planning Sciences, 42*(2), 112–128.

Bettinger, P., Rasheed, K., Maier, F., & Merry, K. (2024). Associations between forest harvest scheduling and artificial intelligence. *International Forestry Review, 26*(4), 387–397.

Bettinger, P., Sessions, J., & Boston, K. (2009). A review of the status and use of validation procedures for heuristics used in forest planning. *Mathematical and Computational Forestry & Natural-Resource Sciences, 1*(1), 26–37.

Bettinger, P., & Zhu, J. (2006). A new heuristic for solving spatially constrained forest planning problems based on mitigation of infeasibilities radiating outward from a forced choice. *Silva Fennica, 40*(2), 315–333.

Boston, K., & Bettinger, P. (1999). An analysis of Monte Carlo integer programming, simulated annealing, and tabu search heuristics for solving spatial harvest scheduling problems. *Forest Science, 45*(2), 292–301.

Cole, B. (1971). What is the Department of Natural Resources? *The Totem, 13*(6-7-8), 1.

Dong, L., Bettinger, P., Liu, Z., & Qin, H. (2015). Spatial forest harvest scheduling for areas involving carbon and timber management goals. *Forests, 6*(4), 1362–1379.

Golden, B. L., & Alt, F. B. (1979). Interval estimation of a global optimum for large combinatorial problems. *Naval Research Logistics Quarterly, 26*(1), 69–77.

Hof, J., & Bevers, M. (2000). Optimal timber harvest scheduling with spatially defined sediment objectives. *Canadian Journal of Forest Research, 30*(9), 1494–1500.

Janová, J., Bödeker, K., Bingham, L., Kindu, M., & Knoke, T. (2024). The role of validation in optimization models for forest management. *Annals of Forest Science, 81*, Article 19.

Lecture, B. (2002). *Draft comparison of the Sessions model with basin planning: An Astoria District case study*. Oregon Department of Forestry.

Los, M., & Lardinois, C. (1982). Combinatorial programming, statistical optimization and the optimal transportation network problem. *Transportation Research—Part B, 16B*(2), 89–124.

Marmorek, D. R., Murray, C., Droessler, T., Dunsworth, G., Monserud, R., Kiester, R., & Northway, S. (2006). *Scientific peer review of H&H model project, review results*. ESSA Technologies Ltd.

Mason, Bruce, & Girard, Inc. (2013). *State of Montana Department of Fish, Wildlife and Parks 2013 forest inventory and sustained yield calculation*. Mason, Bruce & Girard, Inc.

Stander, H., Rasmussen, M., Greer, K., & Baribault, T. (2018). *Minnesota Department of Natural Resources sustainable timber harvest analysis. Phase 2 final report*. Mason, Bruce & Girard, Inc.

State of Montana. (2023). *Montana Code Annotated 2023. Title 87. Fish and wildlife. Chapter 1. Organization and operation. Part 6. Finance. Forest management plan—sustainable yield study required—definition*. Montana State Legislature. Retrieved May 7, 2024, from https://leg.mt.gov/bills/mca/title_0870/chapter_0010/part_0060/section_0220/0870-0010-0060-0220.html

Talbi, E.-G. (2009). *Metaheuristics. From design to implementation*. John Wiley & Sons.

Turing, A. M. (1950). Computing machinery and intelligence. *Mind, 59*(236), 433–460.

U.S. Department of Commerce. (2024a). *QuickFacts Linn County, Oregon*. U.S. Department of Commerce, Census Bureau. Retrieved May 6, 2024, from https://www.census.gov/quickfacts/linncountyoregon

U.S. Department of Commerce. (2024b). *QuickFacts Skagit County, Washington*. U.S. Department of Commerce, Census Bureau. Retrieved May 13, 2024, from https://www.census.gov/quickfacts/fact/table/skagitcountywashington

Washington Department of Natural Resources. (1973). ... Operations research ... *The Totem*, 15(9), 6.

Winograd, T. (1975). Artificial intelligence and language comprehension. In *Artificial intelligence and language comprehension* (pp. 1–25). U.S. Department of Health, Education and Welfare, National Institute of Education.

Zengin, H., Asan, Ü., Destan, S., Ünal, M. E., Yeşil, A., Bettinger, P., & Değermenci, A. S. (2015). Modeling harvest scheduling in multifunctional planning of forests for longterm water yield optimization. *Natural Resource Modeling, 28*(1), 59–85.

Appendix A

Principles of Forest Harvest Scheduling

Seven principles have been offered in this text on forest harvest scheduling. They represent ideas by the author of the text that seem central to the philosophy of forest harvest scheduling. Certainly, they are open to debate.

The*Principle of Scale* (Chap. 1)

The sum of a set of stand-level optimal management decisions does not result in a forest-level optimal management plan; further, the sum of a set of forest-level optimal management plans does not result in a landscape-level optimal course of action.

The *Principle of Optimism* (Chap. 2)

A harvest scheduling effort that utilizes strata to represent land areas, or that is subject to divisibility of decision-variable assignments, can be "worked out" in real life by a field forester when a management plan is implemented.

The *Principle of Effort* (Chap. 3)

A harvest scheduling effort that is relatively easy to implement by a planner (e.g., using linear programming or binary search, recognizing land areas as strata, and avoiding spatial constraints) for some model system (a forest) may be relatively difficult to implement in real life by a field forester.

Conversely, a harvest scheduling effort that is relatively difficult to implement by a planner (e.g., using something other than linear programming or binary search, recognizing areas as individual stands, and incorporating spatial constraints) for some model system (a forest) may be relatively easy to implement in real life by a field forester.

The *Principle of Standards* (Chap. 6)

In moving around a solution space, when solution adjustment standards are too high (disallowing a change to a current feasible solution because the objective function value does not improve or because only a small decline in the objective function value is allowed) a search process may become stuck at a local optimum.

P. Bettinger, *Forest Harvest Scheduling*, Springer Texts in Business and Economics, https://doi.org/10.1007/978-3-031-89432-9

The *Principle of Bias* (Chap. 7)

In a forest harvest scheduling process, bias in the selection of potential moves, or bias in the acceptance of potential moves, is necessary to produce higher quality solutions to complex problems.

The *Principle of Evolving Standards* (Chap. 8)

As the criteria within a heuristic for accepting inferior, feasible solutions tighten, more potential moves (changes to the current solution) are rejected, resulting in an extended time to termination of the search process.

The *Principle of Rationality* (Chap. 10)

The need to make rational integer decisions when moving throughout a solution space may require substantially greater computational work than similar efforts based on irrational, random decisions.

Appendix B

Exam

As one develops an understanding of the various ways in which forest harvest scheduling problems are addressed, the similarities, differences, advantages, and disadvantages of each become clear. The methods employed, for example, by linear programming in seeking the optimal solution to a forestry planning problem are much different from the methods employed by threshold accepting or tabu search. Similarly, threshold accepting and genetic algorithm heuristics are much different in their approaches to solving a problem than a tabu search heuristic. This set of 50 questions represents an examination of the knowledge that might have been gained through engagement with this book. Coupled with real-world experience, a person very familiar with the various methods for approaching optimization in forest harvest scheduling should be able to answer correctly most, if not all, of the questions posed below.

1. **The notion that a strata-based forest harvest schedule will be adjusted by field foresters as it is implemented is suggested in the Principle of ___.**
   (a) Effort
   (b) Optimism
   (c) Scale
   (d) Bias
2. **During the crossover phase in a basic implementation of a genetic algorithm, the parents that are used to form the children are ___.**
   (a) Deconstructed
   (b) Dissolved
   (c) Split and recombined
   (d) Killed

P. Bettinger, *Forest Harvest Scheduling*, Springer Texts in Business and Economics,
https://doi.org/10.1007/978-3-031-89432-9

3. **The notion that the selection and acceptance of potential moves during a search process needs to be non-random in order to locate higher quality solutions to complex problems is suggested in the Principle of ___.**

   (a) Effort
   (b) Optimism
   (c) Scale
   (d) Bias

4. **Although generally a very good heuristic, one of the main problems in using the basic implementation of tabu search for modern timber harvest scheduling, is that ___.**

   (a) Many of the child solutions produced can be infeasible
   (b) The parents are difficult to select
   (c) The tabu state assumed may result in cycling of solutions
   (d) The threshold is often violated

5. **In its general implementation, tabu search is a ___.**

   (a) Point-based, deterministic
   (b) Population-based, deterministic
   (c) Principle-based, stochastic
   (d) Single-agent, stochastic

6. **Basic implementations of threshold accepting and simulated annealing search processes are considered ___.**

   (a) Population-based, stochastic heuristics
   (b) Principle-based, deterministic heuristics
   (c) Point-based, stochastic heuristics
   (d) Single-agent, deterministic heuristics

7. **The process of mutation may be necessary in the basic implementation of this heuristic search process.**

   (a) Tabu search
   (b) Genetic algorithm
   (c) Raindrop method
   (d) Threshold accepting

8. **In contrast to threshold accepting and simulated annealing search processes, the basic implementation of tabu search is ___.**

   (a) A population-based heuristic
   (b) A deterministic heuristic
   (c) An exact algorithm
   (d) A stochastic heuristic

9. **The unit restriction model of timber harvest adjacency ___.**
   (a) Restricts two stands from being scheduled for harvest based on their age
   (b) Restricts two adjacent stands from being scheduled for harvest during the green-up period
   (c) Allows two stands to be harvested during the green-up period as long as their total size is below some maximum size
   (d) Facilitates the development of large areas of old growth forests
10. **The ___ permits two adjacent stands to be scheduled for final harvest during the green-up period, as long as their total size does not exceed the maximum clearcut size.**
    (a) Unit restriction model of adjacency
    (b) Area restriction model of adjacency
    (c) Interior model of adjacency
    (d) Turkish model of adjacency
11. **A green-up period, with respect to forest harvest scheduling, generally refers to a ___.**
    (a) Period of time between two thinnings
    (b) Period of time after a fertilization
    (c) Period of time after a final harvest
    (d) Period of time between thinning and final harvests
12. **Tabu search is generally relatively slow in processing a solution because (in its basic implementation) it has to ___.**
    (a) Assess a large neighborhood of potential solutions before selecting one choice during each iteration of the model
    (b) Develop a large population of parent solutions before applying a crossover and creating child solutions
    (c) Mitigate (repair) adjacency violations caused by the forceful inclusion of a choice (e.g., stand, period of harvest) into a solution
    (d) Resolve many instances of the tabu state by applying the Boltzmann formula to the current threshold
13. **The ___ in tabu search allows the search process to override the tabu state and select a choice (e.g., stand, time period) that is currently tabu.**
    (a) Threshold
    (b) Boltzmann formula
    (c) Aspiration criteria
    (d) Mutation

14. **The notion that a forest harvest schedule that is easy to develop by a planner yet might be difficult to implement by a field forester is suggested in the Principle of ___.**

    (a) Effort
    (b) Optimism
    (c) Scale
    (d) Bias

15. **If the tabu state in tabu search is too short, ___ can occur without you knowing it.**

    (a) The appearance of a cloned population
    (b) The use of a 2-opt move
    (c) Cycling
    (d) Mutation

16. **In simulated annealing, the current ___ is used in part to determine whether to allow the selection of a change (e.g., one stand, different harvest period) to a solution that will lead to an inferior quality objective function value.**

    (a) Tabu state
    (b) Threshold
    (c) Gap variance
    (d) Temperature

17. **In its general implementation, what type of heuristic is a genetic algorithm?**

    (a) Point-based
    (b) Population-based
    (c) Principle-based
    (d) Multiagent

18. **The basis for which heuristic search process arose from work conducted during the Manhattan Project?**

    (a) Threshold accepting
    (b) Raindrop method
    (c) Particle swarm optimization
    (d) Simulated annealing

19. **Prior to the processes of crossover and mutation of child solutions, a basic implementation of a genetic algorithm generally involves the selection of ___ parent solutions.**

    (a) 2
    (b) 10
    (c) 50
    (d) 100

20. **An improvement in the quality of solutions developed by s-metaheuristics can be made with the addition of ___ to the process.**

    (a) Low mutation rates
    (b) 2-opt moves
    (c) Limited crossover
    (d) Cycling

21. **When a constraint violation in a heuristic search is encountered, which of the following will likely be of the least value in addressing it?**

    (a) A reject strategy
    (b) A penalizing strategy
    (c) A repair strategy
    (d) A mutation strategy

22. **With respect to assessing the quality of results generated by a tactical forest harvest scheduling heuristic search process, the lowest level of performance testing involves ___.**

    (a) Computing the estimated global optimum solution
    (b) Comparing the heuristic results with results from mixed integer programming
    (c) Assessing alternative scenarios with the heuristic
    (d) Comparing one set of results from one heuristic to another set of results from another heuristic

23. **With respect to assessing the quality of results generated by a tactical forest harvest scheduling heuristic search process, the highest level of performance testing involves ___.**

    (a) Computing the estimated global optimum solution
    (b) Comparing the heuristic results with results from mixed integer programming
    (c) Assessing alternative scenarios with the heuristic
    (d) Comparing one set of results from one heuristic to another set of results from another heuristic

24. **When using a basic implementation of threshold accepting, the threshold ___ as the program progresses in its attempt to solve a problem.**

    (a) Stays static
    (b) Increases
    (c) Decreases

25. **In contrast to a hill-climbing search algorithm, most of the heuristic search processes examined in this book ___.**

    (a) Can accommodate spatial restrictions in harvest scheduling
    (b) Are easier to develop
    (c) Allow the acceptance of inferior moves as they progress in their attempt to solve a problem
    (d) Are deterministic in nature

26. **In*Linn County v State of Oregon*, what seemed to be the weakest aspect of the forest planning and analysis process in 2000?**

    (a) The algorithm used to schedule harvests (simulated annealing)
    (b) The initial forest inventory
    (c) The modeling of real-life operational concerns
    (d) The formulation of the objective function (goal programming)

27. **Who determined the outcome of*Linn County v State of Oregon*?**

    (a) A panel of three judges
    (b) A jury of twelve random county residents
    (c) The presiding judge
    (d) An independent mediator

28. **The notion that stand-level optimal decisions may not add up to a forest-level optimal decision is suggested in the Principle of ___.**

    (a) Effort
    (b) Optimism
    (c) Scale
    (d) Bias

29. **With respect to forest harvest scheduling using a heuristic method, "reversion" refers to ___.**

    (a) Switching the harvest timing or management prescription assigned to two different stands
    (b) Returning to the initial random solution and beginning the search once again
    (c) Turning around, and conducting an intensive search in the direction from which the search came
    (d) Interrupting the search process and beginning again with the best solution saved in memory

30. **In general, what can we say about heuristic methods when applied to forest harvest scheduling problems?**

    (a) They are all very fast, from a computing point of view.
    (b) They cannot guarantee that the optimal solution to a problem will be found.
    (c) Parameterization, or the specification of search parameters, is easy.
    (d) They are of no value, because linear programming can be used to solve the same problems.

31. **Which of these assumptions is violated when a linear programming problem is altered to become a mixed integer programming problem?**
    (a) Stochasticity
    (b) Certainty
    (c) Divisibility
    (d) Additivity

32. **What is one advantage for using goal programming rather than linear programming for solving a forest management problem?**
    (a) The Simplex Method can be used.
    (b) More than one type of outcome from a plan can be represented in the objective function.
    (c) More than one type of outcome from a plan can be represented by constraints.
    (d) Accounting rows are more compact and efficient.

33. **How would you characterize binary search to describe how it differs from linear programming in addressing a forest harvest scheduling problem?**
    (a) Binary search checks corners of the polytope.
    (b) Binary search simulates harvests one time period at a time.
    (c) Binary search uses a detached coefficient matrix.
    (d) Binary search has accounting rows.

34. **With respect to heuristic search processes, what is a 2-opt move?**
    (a) A doubled reduction in constraints
    (b) The addition of two accounting rows
    (c) The conversion of two decisions from fractions to an integer
    (d) The selection of a change to two decision variables simultaneously

35. **With respect to problem-solving, what is a "hill-climbing search"?**
    (a) The search for money at each iteration of the model
    (b) The search for improbable solutions at each iteration of the model
    (c) The search for only better solutions at each iteration of the model
    (d) The search for average solutions at each iteration of the model

36. **What does the "threshold" represent in threshold accepting?**
    (a) The precipice upon which the search stands before throwing itself off.
    (b) The limit to looking for new solutions, based on the Boltzmann constant.
    (c) The maximum amount by which a new solution can be worse than the previous or best solution.
    (d) The amount a new solution must be different from the previous solution.

37. **What types of adjacency constraints could be used to prevent the scheduling of one final harvest that touches another scheduled final harvest during some pre-defined length of time?**

    (a) Pairwise
    (b) Model 6
    (c) Area evolving
    (d) Modular

38. **Conceptually, what does the basic implementation of the Simplex Method do?**

    (a) It explores solutions through the middle of the solution space.
    (b) It explores solutions in a random fashion.
    (c) It explores solutions around the corners of the polytope.
    (d) It explores solutions one time period at a time.

39. **What cannot be present on the right-hand side of a linear programming constraint?**

    (a) A value over 1,000,000
    (b) A zero
    (c) A variable
    (d) A fractional number

40. **Which of these is one of the main exact solution processes for solving a problem formulated as a MIP model?**

    (a) Binary search
    (b) Greedy search
    (c) Branch and bound
    (d) Threshold accepting

41. **In forestry, binary search has mainly been used in an attempt to "optimize" which of these?**

    (a) Wood harvest volume
    (b) Habitat values
    (c) Stream temperature
    (d) Recreation lease values

42. **If a linear programming constraint has no (0) slack in the final solution to a problem, it is considered ___ to the problem.**

    (a) Irrelevant
    (b) Infeasible
    (c) Temporary
    (d) Binding

43. **A set of linear programming equations is placed in a ___ prior to the application of the Simplex method.**

    (a) Box
    (b) Binary search algorithm
    (c) Object pool
    (d) Detached coefficient matrix

44. **In mixed integer programming problems, one can adjust the ___ to allow the search process to stop within $x\%$ of the relaxed upper bound of the solution (usually the linear programming solution), and thus allow the process to terminate sooner than normal.**

    (a) Number of binary search iterations
    (b) Q-factor
    (c) Optimality tolerance
    (d) Initial threshold

45. **Any feasible solution to a problem can be the optimal solution to the problem, but the optimal solution to a problem is a status generally held by only one feasible solution.**

    (a) True
    (b) False

46. **In contrast to the equations necessary to solve a linear, mixed integer, or goal programming model, a heuristic search process uses ___ and ___ to select potential decisions and assess the potential outcomes and constraint violations.**

    (a) Logic, rules of thumb
    (b) Logic, algebra
    (c) Logic, trigonometry
    (d) Logic, calculus

47. **If a linear programming constraint has a positive dual price in the final solution to a problem, this ___.**

    (a) Places the constraint in a class of infeasibilities
    (b) Suggests that the constraint is of no concern, since the reduced costs of the variables would all have a value of zero (0)
    (c) Indicates how much the objective function value can improve with one additional unit of the right-hand side of the constraint
    (d) Indicates that the variables in the constraint are not included in the final solution to the problem

48. **Unless otherwise constrained, a variable within a linear programming problem will have either ___ or ___ value in the final solution to a problem.**

(a) Negative infinity, positive infinity
(b) Zero, integer
(c) Zero, positive infinity
(d) Negative infinity, zero

49. **In*Concrete School District v State of Washington*, it was argued that the defendants (State of Washington) were ___ in the management and planning of the state forest lands.**

(a) Not recognizing wildlife habitat
(b) Exceeding sustainable harvest rates
(c) Not prudent
(d) Favoring certain tax districts

50. **Although generally a very good heuristic, one of the main problems in using the basic implementation of a genetic algorithm for modern timber harvest scheduling is that ___.**

(a) Many of the child solutions produced can be infeasible
(b) The parents are difficult to select
(c) The tabu state assumed may facilitate cycling of solutions
(d) The threshold is often violated

Appendix C

Data Related to the Contemporary Forest Harvest Scheduling Problem Noted in this Book

See Tables C.1 and C.2.

P. Bettinger, *Forest Harvest Scheduling*, Springer Texts in Business and Economics,
https://doi.org/10.1007/978-3-031-89432-9

Table C.1 Potential objective function values (net discounted revenue) for a forest in southern Arkansas, when scheduling a final harvest to a timber stand and acknowledging reforestation costs

			Time period (year)														
Stand	Age	Acres	1	2	3	4	5	6	7	8	9	10	11	12	13	14	15
1	25	119.535	318,808.87	313,359.15	306,872.63	299,955.48	292,070.00	283,779.09	275,332.00	266,459.46	257,404.97	248,408.31	239,315.76	230,219.04	221,144.39	212,253.69	203,561.19
2	35	121.217	382,679.78	365,787.08	349,143.02	332,769.54	316,922.68	301,607.08	295,103.76	284,286.28	273,417.87	262,713.32	251,923.13	241,374.18	230,967.92	220,719.75	211,008.05
3	25	80.808	230,363.79	226,127.19	221,360.74	216,006.13	210,186.15	204,088.01	197,807.94	191,250.31	184,595.61	177,914.55	171,202.58	164,519.79	157,969.23	151,479.44	145,160.89
4	25	79.506	232,990.08	228,646.97	223,802.04	218,357.37	212,455.55	206,152.33	199,670.47	193,062.36	186,360.24	179,607.39	172,854.38	166,120.50	159,434.83	152,903.46	146,458.01
5	25	40.493	113,239.06	111,240.43	108,901.25	106,194.63	103,454.58	100,447.55	97,279.81	94,098.04	90,813.78	87,554.62	84,239.07	80,978.09	77,734.75	74,563.30	71,468.03
6	24	105.302	252,089.01	248,756.76	244,613.88	239,809.35	234,245.65	228,246.53	221,895.90	215,096.86	208,107.27	201,084.54	193,840.59	186,667.77	179,587.17	172,495.82	165,456.12
7	20	43.249	88,668.61	89,465.69	89,626.99	89,157.94	88,286.77	87,018.19	85,394.63	83,551.25	81,458.63	79,216.88	76,863.12	74,324.27	71,820.49	69,274.03	66,661.22
8	25	119.901	290,692.06	285,805.31	280,137.73	273,977.60	266,846.75	259,476.30	251,742.38	243,578.70	235,369.41	227,021.15	218,592.26	210,271.04	201,949.76	193,774.82	185,778.25
9	15	41.939	54,912.39	59,142.35	62,425.24	64,861.30	66,624.13	67,675.06	68,217.84	68,256.05	67,922.42	67,267.35	66,272.20	65,097.06	63,667.63	62,117.25	60,428.14
10	24	40.062	83,971.46	83,193.90	82,033.50	80,537.99	78,820.22	76,934.38	74,894.93	72,681.51	70,485.81	68,113.43	65,790.11	63,386.15	61,086.74	58,692.17	56,377.98
11	24	79.283	172,029.15	170,138.10	167,796.79	164,654.17	161,074.95	157,166.58	152,957.85	148,408.64	143,898.44	139,144.42	134,204.42	129,468.48	124,604.51	119,806.23	115,009.66
12	14	41.070	44,785.67	49,593.96	53,417.87	56,454.80	58,712.83	60,351.96	61,327.89	61,845.36	61,954.61	61,656.11	61,063.09	60,211.07	59,094.27	57,832.87	56,461.52
13	16	41.370	60,681.20	64,116.87	66,670.45	68,457.54	69,643.22	70,215.77	70,272.01	69,941.17	69,261.03	68,242.05	67,022.81	65,593.42	63,988.92	62,235.13	60,400.63
14	16	40.514	50,627.21	53,977.80	56,436.50	58,325.18	59,498.31	60,246.09	60,494.95	60,420.93	59,914.47	59,181.26	58,253.73	57,114.73	55,804.07	54,422.95	52,879.08
15	5	40.459	(8091.80)	(4897.14)	(475.58)	3268.83	8283.93	12,948.99	17,005.57	20,561.43	23,516.56	25,935.72	27,877.79	29,396.26	30,539.70	31,316.35	31,805.93
16	35	80.812	260,275.63	250,016.62	239,908.67	229,996.75	220,178.60	210,717.94	204,035.98	197,425.19	190,593.37	183,888.88	177,051.76	170,307.11	163,671.59	157,241.60	150,725.14
17	25	57.479	160,778.52	157,980.28	154,597.61	150,909.60	146,873.18	142,567.07	138,124.96	133,573.88	128,969.58	124,256.35	119,602.92	115,021.05	110,404.89	105,899.58	101,495.40
18	25	40.059	109,822.13	107,884.06	105,611.73	103,060.60	100,340.16	97,428.05	94,417.56	91,284.92	88,156.98	84,956.46	81,793.61	78,636.98	75,500.51	72,431.73	69,435.03
19	25	38.837	106,050.66	104,064.22	101,812.88	99,270.72	96,556.65	93,651.08	90,646.25	87,569.46	84,444.80	81,343.44	78,234.63	75,128.36	72,044.78	69,030.07	66,088.55
20	29	76.070	241,385.90	234,208.40	226,942.31	219,489.30	211,794.80	204,144.78	196,363.50	188,699.26	181,183.48	173,731.30	166,450.09	159,198.36	154,652.34	149,898.92	144,978.63
21	32	83.480	285,399.22	275,530.06	265,583.39	255,747.71	245,925.06	236,024.23	226,456.32	217,005.84	207,800.84	202,047.17	195,900.88	189,732.51	183,478.23	177,165.30	170,839.86
22	22	81.800	165,475.23	165,070.59	164,023.62	161,989.63	159,397.54	156,181.98	152,694.61	148,762.91	144,467.52	140,168.07	135,621.00	130,952.59	126,289.95	121,590.86	116,930.37
23	10	80.809	19,287.43	31,141.24	41,998.40	51,526.68	59,600.18	66,366.74	71,746.37	76,000.15	79,255.51	81,537.85	83,052.46	83,813.08	83,994.18	83,671.78	82,914.46
24	26	89.498	202,510.33	199,084.46	194,924.18	190,334.71	185,352.69	180,072.47	174,540.21	168,961.83	163,227.95	157,401.94	151,509.77	145,793.66	140,058.80	134,426.88	128,909.09
25	27	83.082	205,290.92	201,235.02	196,701.49	191,773.21	186,524.28	181,003.28	175,415.75	169,776.58	163,903.97	158,073.35	152,277.00	146,560.18	140,835.15	135,141.68	131,652.22
26	7	39.662	(7202.77)	(4324.13)	516.17	5136.52	9469.26	13,406.89	16,868.41	19,795.56	22,243.39	24,262.10	25,897.43	27,112.20	28,032.10	28,655.74	29,019.67
27	30	119.452	348,225.12	338,326.85	327,801.57	317,005.37	306,407.89	295,463.93	284,642.13	273,776.51	262,946.15	252,331.29	242,094.52	235,514.74	228,920.96	221,880.55	214,704.34
28	26	51.282	136,385.93	133,768.61	130,752.27	127,392.91	123,918.27	120,201.07	116,292.79	112,377.49	108,472.11	104,525.67	100,562.29	96,669.36	92,784.35	89,036.23	85,280.33
29	9	100.478	15,071.70	27,632.34	42,603.45	54,199.42	64,195.97	72,603.85	79,586.92	85,168.58	89,387.58	92,519.44	94,692.77	95,924.15	96,430.09	96,393.65	95,800.82
30	8	37.903	1287.27	6746.71	12,220.45	17,233.04	21,582.37	25,277.23	28,383.82	30,915.05	32,889.31	34,414.03	35,501.02	36,204.00	36,607.04	36,714.33	36,628.92
31	9	61.095	1844.38	10,113.63	17,953.79	24,873.98	30,953.23	36,264.57	40,550.40	44,004.86	46,793.68	48,989.33	50,529.01	51,555.26	52,131.57	52,369.37	52,260.26

Table C.2 Potential wood volume (total tons) for a forest in southern Arkansas, when scheduling a final harvest to a timber stand

Stand	Age	Acres	Time period (year)														
			1	2	3	4	5	6	7	8	9	10	11	12	13	14	15
1	25	119.535	17,882.4	18,575.7	19,233.2	19,878.7	20,476.3	21,050.1	21,611.9	22,137.9	22,639.9	23,130.0	23,596.2	24,038.5	24,456.9	24,863.3	25,257.7
2	35	121.217	19,515.9	19,794.7	20,049.3	20,279.6	20,497.8	20,703.9	21,467.5	21,940.3	22,388.8	22,825.2	23,225.2	23,613.1	23,976.7	24,316.1	24,667.7
3	25	80.808	12,888.9	13,373.7	13,842.4	14,286.9	14,707.1	15,111.1	15,499.0	15,862.6	16,210.1	16,541.4	16,856.5	17,155.5	17,446.4	17,721.2	17,987.9
4	25	79.506	12,903.8	13,388.8	13,857.9	14,303.1	14,724.5	15,122.0	15,503.7	15,869.4	16,219.2	16,553.1	16,871.2	17,173.3	17,459.5	17,737.8	18,000.2
5	25	40.493	6345.3	6588.2	6819.0	7033.6	7248.2	7446.7	7632.9	7815.1	7985.2	8151.2	8305.1	8454.9	8596.7	8734.3	8868.0
6	24	105.302	14,078.9	14,679.1	15,258.3	15,816.4	16,342.9	16,848.3	17,332.7	17,785.5	18,217.2	18,638.5	19,028.1	19,407.2	19,775.7	20,123.2	20,449.6
7	20	43.249	4649.3	4952.0	5241.8	5514.2	5778.1	6028.9	6266.8	6496.0	6712.2	6919.8	7118.8	7300.4	7482.1	7655.1	7815.1
8	25	119.901	16,078.7	16,714.2	17,325.7	17,925.2	18,476.7	19,016.3	19,531.9	20,011.5	20,479.1	20,922.7	21,342.4	21,750.0	22,133.7	22,505.4	22,865.1
9	15	41.939	3271.2	3669.7	4051.3	4416.2	4768.5	5099.8	5418.5	5720.5	6009.9	6286.7	6546.7	6798.3	7033.2	7259.6	7473.5
10	24	40.062	4823.5	5043.8	5252.1	5448.4	5636.7	5817.0	5989.3	6149.5	6309.8	6454.0	6598.2	6730.4	6866.6	6986.8	7107.0
11	24	79.283	9886.6	10,322.6	10,750.8	11,147.2	11,527.7	11,892.5	12,241.3	12,566.4	12,891.4	13,192.7	13,470.2	13,755.6	14,017.2	14,270.9	14,508.8
12	14	41.070	2739.4	3137.7	3519.7	3889.3	4242.5	4583.4	4903.8	5211.8	5507.5	5786.8	6053.7	6308.4	6546.6	6776.6	6998.3
13	16	41.370	3545.4	3917.7	4273.5	4612.8	4939.6	5249.9	5543.6	5824.9	6093.8	6346.2	6590.2	6821.9	7041.2	7248.0	7446.6
14	16	40.514	3135.8	3480.2	3804.3	4120.3	4416.0	4703.7	4975.1	5238.5	5481.5	5716.5	5943.4	6158.1	6360.7	6559.2	6741.5
15	5	40.459	97.1	250.8	497.6	732.3	1080.3	1444.4	1804.5	2164.6	2512.5	2848.3	3172.0	3483.5	3782.9	4066.1	4337.2
16	35	80.812	13,584.5	13,843.1	14,093.6	14,336.0	14,562.3	14,788.6	15,192.7	15,588.6	15,960.4	16,332.1	16,679.6	17,019.0	17,350.3	17,681.7	17,980.7
17	25	57.479	9156.4	9507.0	9834.7	10,150.8	10,449.7	10,731.3	11,001.5	11,260.1	11,507.3	11,737.2	11,961.4	12,179.8	12,381.0	12,576.4	12,766.1
18	25	40.059	6341.3	6581.7	6810.0	7026.3	7234.7	7430.9	7619.2	7795.5	7967.7	8128.0	8284.2	8432.4	8572.6	8708.8	8841.0
19	25	38.837	5580.9	5798.4	6008.1	6206.2	6396.5	6575.1	6746.0	6909.1	7064.5	7215.9	7359.6	7495.5	7623.7	7748.0	7868.4
20	29	76.070	13,152.5	13,510.0	13,860.0	14,194.7	14,506.5	14,810.8	15,092.3	15,366.1	15,632.4	15,883.4	16,126.8	16,347.4	16,819.1	17,267.9	17,693.9
21	32	83.480	16,019.8	16,370.4	16,704.3	17,029.9	17,338.8	17,622.6	17,906.5	18,173.6	18,432.4	18,966.7	19,467.5	19,960.1	20,435.9	20,895.0	21,337.5
22	22	81.800	8973.5	9456.1	9930.5	10,372.2	10,797.6	11,198.4	11,591.1	11,959.2	12,302.7	12,646.3	12,965.3	13,268.0	13,562.4	13,840.6	14,110.5
23	10	80.809	2012.1	2739.4	3490.9	4242.5	4977.8	5697.0	6383.9	7046.5	7684.9	8291.0	8872.8	9422.3	9947.6	10,448.6	10,925.4
24	26	89.498	11,661.6	12,109.1	12,529.7	12,932.5	13,317.3	13,684.2	14,033.3	14,373.4	14,695.6	14,999.9	15,286.3	15,572.7	15,841.1	16,100.7	16,351.3
25	27	83.082	11,905.7	12,329.4	12,736.5	13,127.0	13,500.8	13,858.1	14,207.0	14,547.7	14,863.4	15,170.8	15,469.9	15,760.7	16,034.8	16,292.4	16,790.9
26	7	39.662	130.9	269.7	543.4	836.9	1146.2	1463.5	1780.8	2090.2	2391.6	2685.1	2970.7	3240.4	3502.2	3752.0	3990.0
27	30	119.452	19,960.4	20,509.9	21,023.6	21,513.3	22,003.1	22,457.0	22,898.9	23,317.0	23,711.2	24,093.5	24,475.7	25,192.4	25,909.1	26,578.1	27,223.1
28	26	51.282	7666.7	7948.7	8215.4	8466.7	8712.8	8943.6	9159.0	9369.2	9574.3	9769.2	9953.8	10,133.3	10,302.6	10,471.8	10,625.6
29	9	100.478	2059.8	2813.4	3798.1	4682.3	5556.4	6410.5	7244.5	8048.3	8811.9	9545.4	10,248.8	10,911.9	11,544.9	12,157.8	12,740.6

(continued)

Table C.2 (continued)

Stand	Age	Acres	Time period (year)														
			1	2	3	4	5	6	7	8	9	10	11	12	13	14	15
30	8	37.903	549.6	860.4	1209.1	1569.2	1925.5	2274.2	2615.3	2945.1	3259.7	3562.9	3850.9	4123.8	4385.4	4631.7	4870.5
31	9	61.095	892.0	1362.4	1863.4	2364.4	2865.4	3366.3	3842.9	4301.1	4747.1	5180.9	5590.2	5981.2	6353.9	6714.3	7056.5

GPSR Compliance
The European Union's (EU) General Product Safety Regulation (GPSR) is a set of rules that requires consumer products to be safe and our obligations to ensure this.

If you have any concerns about our products, you can contact us on

ProductSafety@springernature.com

In case Publisher is established outside the EU, the EU authorized representative is:

Springer Nature Customer Service Center GmbH
Europaplatz 3
69115 Heidelberg, Germany

www.ingramcontent.com/pod-product-compliance
Lightning Source LLC
Chambersburg PA
CBHW071142040926
50963CB00003B/147
* 9 7 8 3 0 3 1 8 9 4 3 1 2 *